MINERAL PROCESSING TECHNIQUES

Basics and Related Issues

G. S. Ramakrishna Rao

B.Sc. (Hons.), M.Sc., M.S. (M.I.T., USA)

Zorba Publishers

Published in India by Zorba Publishers, 2014

Website: www.zorbapublishers.com
Email: info@zorbapublishers.com

ISBN 978-93-81239-544

Zorba Publishers Pvt. Ltd.
Gurgaon, INDIA

Cover design Qualcom

Printed in India

Dedicated to

Professor A. M. Gaudin

For his unforgettable teaching, encouragement,
and advise to promote mineral industry on my return to India

Acknowledgements

I am indebted to my late professor A. M. Gaudin for his guidance at MIT, Cambridge, USA, and his advice to develop mineral industry in India. I was fortunate to learn while working with him both at MIT as well as my association with him while he was advisor to White Pine Copper Company, White Pine, Michigan, USA.

It was late Professor C. Mahadevan, former Head of the Geology Department, Andhra University, who encouraged me to join the newly opened course in ore dressing at the graduate level in 1954. As there was no text book at that time in mineral processing practices, only brief lecture notes were made available to graduate students. I could overcome the knowledge gap, besides others, from the valuable teaching of Dr G. S. R. Narasimhamurty, former Head, Ore Dressing Department of Andhra University.

At the end of over 50 years of service in mineral engineering field, my children, my wife, colleagues, and a few past students of the Indian School of Mines, Dhanbad, encouraged me to write a simple book useful to the students pursuing mineral processing while preparing for graduate studies. An attempt is made to fulfil this suggestion. The present book is prepared to meet the needs of the students of ore dressing, applied geology, mining, and metallurgy besides providing basic knowledge on related topics to mine owners, entrepreneurs, and industry.

This book is based on 'hands on experience' gained by the author while working with late Sri P.I.A. Narayanan, the father of ore dressing in India then, in testing different ores at the Mineral Beneficiation Pilot Plant of National Metallurgical Laboratory, Jamshedpur. Further, the information included in this book is based on the personal expertise and knowledge gained during the course of my long association with the National Mineral Development Corporation Limited, Hyderabad; the UNIDO, Jos, Nigeria; and the Mineral Sales Private Ltd., Hospet. Grateful thanks are therefore due to the Director, National Metallurgical Laboratory, Jamshedpur; the Chairman cum Managing Director, National Mineral Development Corporation Limited Hyderabad; Secretary General of the United Nations and the Chairman, Mineral Sales Pvt. Ltd., Hospet.

I also owe my sincere thanks to the Controller General, Indian Bureau of Mines, as well as to the Director General, Geological Survey of India, for letting me reproduce the Indian mineral reserves data published by them earlier.

My sincere thanks are due to my daughter, Dr G. Sandhya Devi for editing this book with good suggestions for its improvement. This leads to better understanding the subject even by the beginners unconnected with the field of mineral industry.

I could not have completed this book without active encouragement of my wife, G. Radha Devi and valuable suggestions for its improvement by my sons G. Ravi Kumar and G. Madhukar; Sri P. R. Tripathi, former CMD, NMDC Ltd.; Sri N. K. Nanda, Director (technical), NMDC Limited and Dr H. Sarvothaman, Former DDG, Geological Survey of India.

I am indebted to several authors, editors, and publishers for letting me include some pertinent information, published earlier by them, for the completeness of the book. I am particularly thankful to the following publishers, for using their information published long back, though specific permission was not formally sought for and obtained for including the same.

1. McGraw-Hill
2. Sir Isaac Pitman & Sons
3. John Wiley
   Chapman & Hall
4. Elsevier Science & Technology Books
5. United Steel Corporation
6. *Mining Engineers' Journal*

Though acknowledgements have been made for the source at the respective places in the book where specific information is included, still, I beg to be forgiven for any inadvertent omission of acknowledgement for any particular source, author, editor, or publisher of information in the overall country's interests and development of Indian mineral industry. I also owe my sincere thanks to those authors who published pictures of rocks and minerals in the internet. In anticipation of their due approval, six pictures of minerals are included for the benefit of students seeking knowledge.

Last, but not the least, I couldn't have completed the book without the active support from my son-in-law, N. V. Ramana, daughters-in-law, Anuradha and Vijayalaxmi, and grand-children, Sagar, Nikhila, Pooja and Nidhi.

G. S. Ramakrishna Rao
B.Sc. (Hons.), M.Sc., M.S. (M.I.T., USA)
October 2014

Table of Contents

List of Figures

List of Tables

List of Photos

Foreword for Students

Mr G. S. Ramakrishna Rao has attempted this title for the benefit of students of mineral engineering in a very simple and lucid manner. Since Mr Ramakrishna Rao has, apart from high academic qualification, vast experience in operation of number of pilot plants as well as commercial scale plants, he has tried to share his vast knowledge through this treatise.

I find almost all the metallic, non-metallic, and industrial minerals have been covered, which would give a good preliminary understanding to any student of mineral dressing. The book does not claim to be the final solution to different minerals but gives an idea of fundamental processes that are required to be studied and applied before taking up the problem of upgradation of the minerals and reclaim useful values from waste rock.

I find the book to be useful and hope it would serve its purpose.

P.R.TRIPATHI
AISM (Mining Engg), BSc (Hons) Mining Engg
FCC (Coal & Metal), FIE, FAIMA
Former CMD NMDC Ltd.

Review

Dear Mr Ramakrishna Rao,

The idea of transferring your thoughts, knowledge, and experience in the form of a book is very good. Books are the best friends, and it is a highly useful book for young mineral engineers of India, educational institutions, mineral-based industries, and Earth Science Library.

This book describes the major Indian deposits and beneficiation processes adopted in India and abroad. I feel that information about mineral beneficiation is covered to a large extent. Only thing is to emphasize on genesis and mineralogy required, which will help to know the formations of mineral and gangue at various conditions and temperatures. Also it helps to select suitable process for separation of valuable concentrate and the gangue through proper techniques.

With all the best wishes

Narendrakumar A. Baldota
Chairman & Managing Director
Mineral Sales Private Limited
24 August 2013

BOOK REVIEW BY SHRI NARENDRA KUMAR NANDA, DIRECTOR (TECHNICAL), NMDC LIMITED

India's crude steel production is around 81.2 MT and Govt. of India is gearing up to achieve a target of 300 MT crude steel production by 2025. To meet the increased steel demand, India also needs to increase its Mineral production capacity especially Iron Ore, which was around 150 MT in the year 2013. The Iron Ore consumption is expected to increase rapidly considering the increasing demand of steel in the recent future. India's Iron Ore production is facing challenges because of the limited identified Iron ore reserves. Identification of new proven reserves, obtaining of new mining leases, starting mining activities, because of the clearance issues etc. It has therefore become necessary to shift the focus on utilisation of Iron Ore fines, low grade iron ores through beneficiation route.

This book largely covering topics relating to application of mineral processing techniques will serve as a handbook for practicing engineers. The book has covered topics related to various beneficiation process of low grade / high grade Iron Ore and other minerals. The topics related to BHJ, BHQ & BMQ are also useful for the students, as these subjects are generally not covered in other academic books. The book also provides basic information on ores of alkaline earth metals and their beneficiation, diamonds, base metallic ores extraction of precious metals like gold, silver, platinum etc. The author in this book has tried to share his practical lifetime experience, which is well reflected. Other basic information such as environmental issues, preparation of flowsheets, estimation of water and power requirement, equipment & capital costs etc. are also very useful. No doubt, the book will be useful for the students and this will also be very much useful for the new comers in the mineral industry/sector and also for the professionals associated with the mineral industry.

The book should be read from the perspective of trying to understand and use it to solve work problems, much as a student must read the book in order to undertake homework assignments and be tested on its content and knowledge imparted.

I like to congratulate Shri G. S. Ramakrishna Rao, the author of the book and all those who are directly or indirectly associated with the publication of the book, thus providing a gateway for sharing knowledge.

(Narendra Kumar Nanda)

Preface

This book, like lecture notes prepared from several standard text books for teaching, may serve as a basic guide for graduate students of ore dressing, metallurgy, applied geology, and mining pursuing mineral processing.

This book includes basic elements of current practices of ore processing adopted mostly in India on a commercial scale, which makes the book suitable for understanding the subject in a very simple and concise way. For reference purposes, a few processing techniques adopted abroad are also included.

An attempt is made to include the reserves of each economic ore available in India in order to give some basic information for professionals engaged in mineral engineering industry.

For those in the mineral industry who are interested to know the application aspects, the following aspects, by way of examples, have also been included:

- Scope of process flowsheet, material balance flowsheet, drawing of equipment specifications, equipment flowsheet, and so on.
- Milestones to be kept in mind while setting up a commercial beneficiation plant.
- Estimation of water requirement, power requirement, and choosing the appropriate site for a process plant.
- Basis for estimation of equipment and capital costs of the plant.
- Estimation of production costs and profitability.
- Issues connected with environment and mining affecting the smooth operation of the plant.
- Basic conversion factors and common formulae used in mineral engineering.
- Genesis of rocks and mineral deposits.
- Characteristics of common economic minerals.
- Opening sizes of standard test sieves.

As the industry is dynamic with incorporating improved techniques/latest innovations, making modifications to processes to cope with the changes in the quality of ore being treated/mined from lower depths, the professional is strongly advised to refer to the latest journals, websites, latest text, and other reference books for more information on theory, latest plant practices, operating parameters, current results, and techno economics.

Chapter 1

Introduction: Basic Principles

India has rich resources of various minerals such as bauxite, mica, monazite, iron ore, barytes etc. However, due to increasing demand for these ores both in India and international markets like China, Japan, over exploitation of some of these high grade ores over the past decade has caused fast depletion of these resources. This has, created great concern and the realization that it is imperative to preserve these high grade ores for meeting domestic requirements of the future. Conservation of these high grade resources for posterity is also recognized by the Government of India. Maximum utilization of the available medium and low grade ores as well as the existing tailing dumps after concentration with the state-of-the-art technology is absolutely necessary. Everyone connected with the mineral industry, be it working industrialists, professionals, experts or students have to strictly follow guidelines set out by the Government of India for mining, ore processing and metal extraction. Moreover, ore extraction has to be carried out in an eco-friendly manner, which calls for utilizing innovative technology to enable optimal utilization of the available mineral wealth, including low grade ores, and ensuring a pollution free environment. Therefore, given the magnitude and gravity of the need to protect environment, specifically in the context of winning minerals, metals, building materials from the mother earth, it is expected of each and every one of us in the mineral industry to do our utmost towards preserving and protecting our planet for posterity.

To start with, let us review the thumb rules/guidelines of mineral processing. Let us suppose an ore containing two minerals, say 'X' and 'Y' needs to be processed in order to separate these two minerals. Then, it is necessary to know the following:

- *Identify the two minerals to know their physical, chemical, and other properties:* This step helps to identify/select the best property that can be utilized to separate the desired mineral from the other associated minerals of the ore. It is necessary to find out in which particular property, like specific gravity, magnetic property, colour, hardness, size, surface properties, electrical conductivity etc., the two minerals in the ore differ from each other.
- *Tenor or grade:* Tenor is the metal or desired mineral content in the ore. Separation of the desired minerals from the ore on a commercial scale should be attempted only if it is techno-commercially viable. For example, presently gold ore is mined and extracted only if it contains about one gram gold/tonne at its existing market price of say Rs 3000/gram. The separation is not economically viable with the fall in price of gold or its incidence in the ore/rock.
- *Extent of minimum crushing and grinding required to achieve liberation of 'X' from 'Y':* This is the size at which majority of grains of "X" (say 80%) are loosened/set free from grains of "Y" upon crushing and grinding. These free grains can be identified as 'X' or 'Y'. The few interlocked grains contain both 'X' and 'Y'. This size reduction is a prerequisite step for their ultimate separation producing a saleable concentrate. Achieving liberation at a coarser size through crushing and grinding is desirable as it saves energy and minimises production of undesirable fines.
- *Grain size of the two minerals in free-state and the nature and extent of interlocking of the associated minerals:* Knowing the grain size and degree of interlocking will determine the processing route for their separation. For example, Jigging is used for separation of coarse particles and hydro cyclones for fine particles.
- *Simplicity/complexity of processing route is primarily based on the value of the desired element/mineral:* Processing and extraction of high value minerals such as gold, platinum, diamonds, nickel, tungsten, uranium etc. can be quite complex, particularly for extracting almost the entire quantity of the aforesaid precious metals

from their corresponding ore. On the other hand, the process route should be as simple as possible, to extract low-value ores like iron ore, limestone, coal, barytes, silica sand etc. profitably.

- *Choosing an optimum process based on actual testing rather than blindly adopting methods already in use or available data base:* No two ores are alike. The same mine may contain different combination of ores at different areas on the same bench and the ore quality changes with depth. Hence it is absolutely necessary to get a representative sample of the ore deposit tested by a reputed testing agency to know the grade of the concentrate, percentage yield, and recovery of the desired mineral obtainable through the recommended process under optimum operating conditions.
- *Ore testing agencies:* Large mining organizations like SAIL, NMDC, KIOCL, TISCO etc. have their own in-house R&D laboratories, including a pilot plant, to repeat ore testing whenever necessary for finding the modified operating conditions needed to obtain the desired end products and maintain the profitability. There are also other excellent National Laboratories of CSIR, like NML, Jamshedpur/Chennai; RRL, Bhubaneswar/Bhopal; and the Indian Bureau of Mines, Nagpur/Bangalore etc. These laboratories have equally good ore testing facilities, including pilot plants, for continuous ore testing. Being Government Organizations, testing charges of these laboratories are relatively lower but heavily booked. Testing is time consuming. Due to pending assignments, it may take a few months to get the ore tested by these Institutes and obtain the test results. But report of these testing agencies may help in enhancing the confidence level for putting up a commercial plant, estimation of capital, operating costs, and profitability, getting sanction of loans from the banks and technical assistance of the testing institute, or CSIR if required, in setting up the production plant. It is therefore desirable for a major mining organization to have its own testing facilities for frequent testing, and save time in taking corrective actions for keeping up production schedule/profitability but setting up an in-house R & D Centre will

be quite expensive. The choice between having an in-house testing laboratory involving a large capital investment and outsourcing ore testing by an Indian or foreign laboratory involving time delays is left to one's discretion, based on urgent need besides confidence in test results and investment considerations.

- *Need for classified feed for effective gravity concentration:* After crushing and grinding ore to the liberation size, closely sized feed such as -10 mesh + 20 mesh fraction is treated by jigging and -20 + 48 mesh fraction by tabling or spiral concentration. Similarly, -48 + 100 mesh classified feed particles having more or less same settling velocities (coarser, light particles along with finer, heavy grains) are treated by tabling or spirals or hydro cyclones for economic recovery.
- *Specifications of the saleable product:* All efforts have to be made for meeting the market specifications of the final saleable product (e.g., size, purity/grade, physical and metallurgical properties, moisture content, etc.).
- *Utilities and consumables for ore testing:* It is necessary to adopt the simplest possible flowsheet, with the minimum use of resources like power, water, and chemicals if any - for the production of the desired quality and quantity of the product. Processing the ore by the simplest feasible process route will enhance profitability.
- *Separation of ore particles* in *the coarsest size soon after their liberation:* The desired mineral should always be recovered in the coarsest possible size soon after it is liberated from the associated gangue. This is necessary to reduce grinding/operational costs and achieve better grade, better sale price with higher metal recovery. Over grinding converts the desired mineral into slimes, which poses difficulty for its separation/recovery, leading to more financial losses.
- *Rejects:* The rejects or unwanted minerals should be contained in a tailing dam/waste dumps to minimize environmental pollution, water recovery and reprocessing the same on increased market demand by the more advanced processing techniques. Attempts

should be made for finding commercial use of the rejects for making bricks, pavement tiles, carpeting roads etc. to enhance profitability.

- *Requirement for commercial exploitation:* Commercial exploitation of the ore should be undertaken only if the process is techno-economically viable, availability of market demand for the finished product, generates employment, improves overall economy and living conditions of the local residents, and the venture is eco friendly.
- *Economics:* Commercial production is undertaken preferably with minimum capital, low operating costs, and getting reasonable profitability to the investors.
- A general ore processing flowsheet including the major unit operations is shown in Fig. 1.1.

Common processing methods used for achieving the separation of the desired mineral from the other unwanted minerals in the ore is shown in Table 1.1.

6 Mineral Processing Techniques

Figure 1.1: General ore processing flowsheet

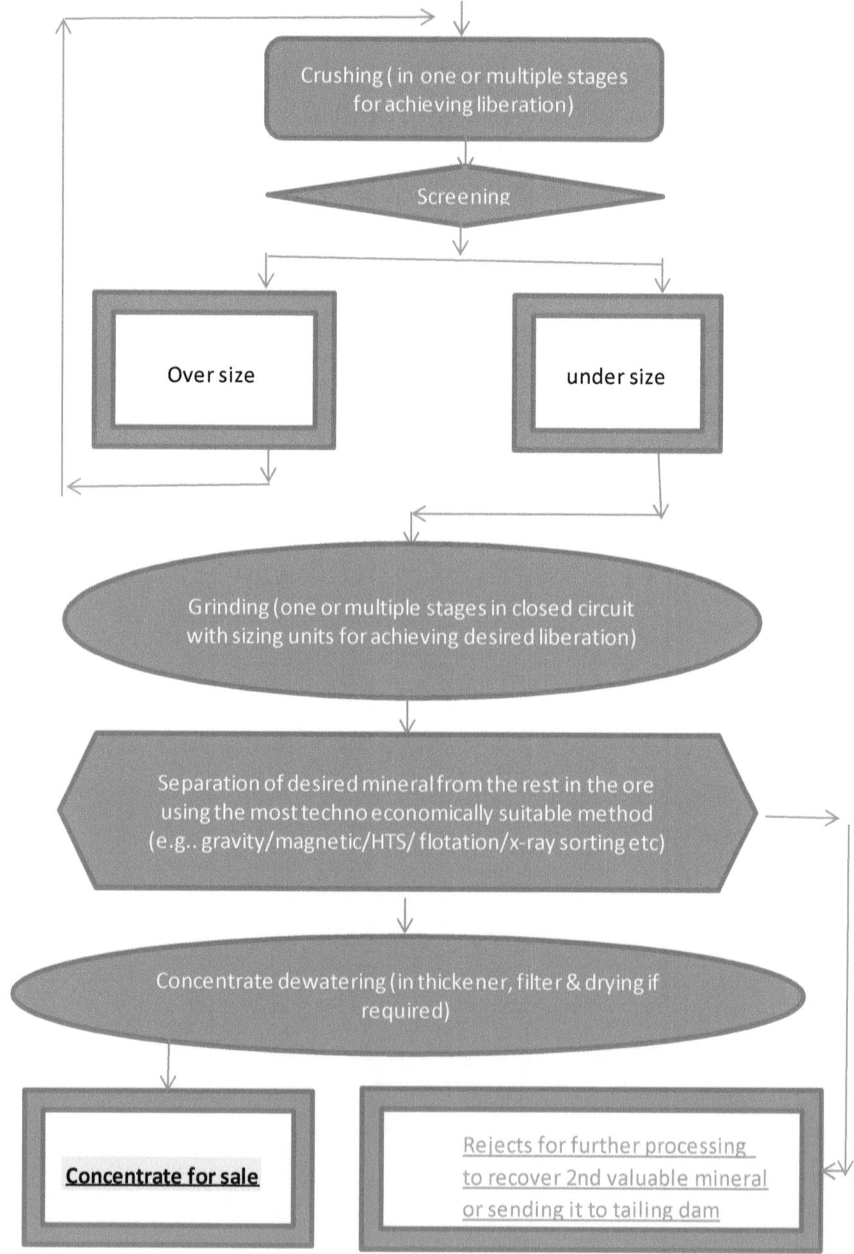

Table 1.1: Common processing methods used for achieving separation of the desired mineral from the other associated minerals in the ore

Ore/deposit	Ore minerals	Gangue	Simpler ores	Complex ore
High grade iron ores of Bailadila, Kiriburu, Bellary, Hospet, Goa, etc.	Hematite (high grade ores)	Sili-cates, clay/ laterite	Wet screening, classifi-cation, cycloning	Scrubbing, wet screening, jigging, classification, cycloning
Low grade magnetite ores of Kudremukh, Bababudan, Ongole, Kodachadri, etc.	Magne-tite, hematite	Quartz, silicates	Low intensity magnetic separation, gravity concen-tration (cyclones/ spirals)	Low, medium, and high intensity magnetic separation, flotation
Hard and low grade iron ores like BHQ, BHJ, taconites, Itaberites etc.	Hematite, magnetite	Quartz, silicates	Medium intensity magnetic separation followed by high intensity magnetic separation	Medium and high intensity or high gradient magnetic separation followed by cationic flotation after regrinding, if required
Limestone, dolomite of Birmitrapur	Calcite, dolomite	Sili-cates	Anionic flotation	

Magnesite	Magnesite	Sili-cates	Anionic flotation	
Coking and non-coking Coal of Jharia, Raniganj	Coal	Slate, shale, and silicates	Gravity (Chance cone, cyclones)	
Diamondi-ferous Kimberlite tuff of Panna	Diamonds	Quartz , silicates	Heavy media separation, X-ray sorting, grease tables	
Manganese ores of MP	Pyrolu-site, psiloma-lene	Sili-cates, ferrugi-nous mine-rals	Gravity methods like jigging, tabling etc.	Gravity methods, reduction roast followed by magnetic separation
Chromite	Chromite	Sili-cates	Hand sorting/ gravity methods	Jigging, tabling, and magnetic separation
Copper lead zinc ores of Bihar, Rajasthan	Chalcopy-rite, galena, sphalerite	Pyrite, silicates	Selective/ differential flotation	

Copper ores of Bihar (Ghatsila) & Rajasthan (Khetri)	Chalcopyrite; chalcocite with traces of gold , silver, molybdenum, nickel etc.	Pyrite, silicates	Selective / differential flotation	Recovery of precious metals like gold/silver during smelting
Precious metals of Kolar	Gold	Quartz, silicates	Gravity, panning, multi gravity separation, amalgamation, Cyanidation, electrolysis, and precipitation	Amalgamation, cyanidation, precipitation, electrolysis of enriched concentrate/ hydro-metallurgy
Nickel ores of Sukinda, Orissa			Gravity or magnetic separation	Gravity methods, flotation
Bauxite			Bayer's process (hydro-metallurgy)	
Barytes			Crushing , grinding, sizing	

Silica sand			Crushing, Raymond mills, sizing (Derrick screens)	
Beach sands	Ilmenite, monazite, garnet, rutile, zircon	Sillima-nite, quartz	Low , medium, and high intensity magnetic separation followed by high tension separation of sized and dried product and tabling	
Uranium minerals of Jaduguda	Monazite, uraninite,		Gravity, hydrometa-llurgy, ion exchange	
Marbles and granites	Calcite		Selective mining, cutting, and polishing	

(From images of hematite-report images)/Images of minerals (Internet)

(From images of Diamond-report images)/Images of minerals (Internet)

From images of copper ore-report images (Internet/buy/sell/rent any thing)

From images of lead ore-report images (Internet/buy/sell/rent anything)

From images of sphalerite-report images (internet/buy/sell/rent anything)

From images of gold ore-report images (Internet/buy, sell, rent anything)

Chapter 2

Need for Beneficiation of Various Ores

2.1 Beneficiation

'Beneficiation' is the process used for improving the quality of low grade ores or rocks. Beneficiation is required as low grade ores or rocks cannot be used as such in their original state. The process is utilized for producing a richer concentrate for meeting the requirements/ specifications of the ultimate user/buyer. Beneficiation is an important separation process that is specifically used for improving an ore's grade and includes besides crushing, sizing, and grinding, an additional essential separation step of concentration. Without upgradation by beneficiation, the key process involved in ore enrichment involving physical methods of separation using differences in the physical properties of the associated minerals, ores such as gold, diamonds, low grade iron ores (BHJ) etc. remain as just rocks and will not be of any value.

2.2 Processing

'Processing' is used for high grade ores such as iron ores of Bailadila sector as well as Bellary-Hospet sector in India, and the term 'processing' essentially denotes a simpler processes such as crushing, sizing, classification without any need for its significant enrichment. It is carried out only for meeting just the physical requirements/ specifications of the ore as required by the ultimate user.

2.3 Separation

'Separation' is used for processing high grade pre liberated (pre liberated due to natural weathering over centuries) sandy ores such as beach sands containing an admixture of pure minerals of ilmenite, rutile, zircon, monazite, magnetite, sillimanite, and silica. Sandy ores are available at Visakhapatnam, Kerala, and other sea coasts. In this case, crushing and grinding operations are not required as pure minerals are already present as an admixture along with gangue minerals like silica or silicates. The required separation technology involves recovery of these minerals into saleable products using differences in their physical properties such as magnetic property, conductivity, and specific gravity.

2.4 Extraction

'Extraction' is the process carried out for enriching ores such as bauxite, very low grade platinum, gold, uranium ores. They cannot be enriched by physical methods of separation alone. The metal and/other valuable ingredients can be recovered from the crushed and sized ore by using other methods such as pyro metallurgy, chemical extraction, ion exchange, electrolysis etc. Hence, the term 'extraction' refers to enrichment of ores by chemical dissolution or heating in lieu of other physical methods of separation.

Iron ores

- *High grade iron ores:* High grade iron ores of Bailadila region are richer than 63% Fe. The ore from these mines meets the chemical specifications of all the steel plants. Processing is required only to meet the size specifications of the buyer (to meet buyer's size requirement such as -100 + 10 mm or -40 +10 mm or -30 + 6 mm). In a few mines, high grade hard iron ore occurs along with high grade but soft and fine iron ore called blue dust or specular hematite. During crushing, sizing, and classification, a very small quantity of fines (<10%), unsuitable for marketing, are rejected, which in turn lead to improvement in one or more of the metallurgical specifications of the ore as required by different end-users, such as steel plants/and importers. These improvements include higher tumbler/shatter and reducibility indices as well as lower thermal degradation and decrepitating properties. As the reserves of high grade iron ores are fast depleting, there is an urgent need to preserve/conserve them keeping in view the requirements of future generations. Their use should be restricted to sweeten the export of available low grade ores or meeting the requirements of local steel plants.

Photo 2.1 is an aerial photograph that shows mining operations carried out at Bailadila iron ore mine.

Photo 2.1: Aerial photograph of Bailadila iron ore mine

Courtesy: NMDC

Medium grade ores: The medium grade ores are richer than 56% Fe. These are available from mines such as Kiriburu, Meghataburu, Barsua, and Goa located in states of Bihar, Orissa, Goa. The object of processing the ore involves essentially reducing clay/alumina selectively and improving the silica to alumina ratio. This step may also improve iron content by 1 or 2%. Reduction of alumina besides improving its iron content and silica to alumina ratio of the ore reduce coke and flux consumption and smelting costs in iron and steel making and increase steel productivity. -30 +6 mm sized lumps are used directly for smelting in blast furnaces and direct reduction plants (DR plants). The -6 mm + 100 mesh resulting fines are agglomerated along with coke fines, flux, steel plant wastes etc. through sintering technique, which is carried out separately prior to smelting of the sinter in blast furnace.

- *Low grade banded magnetite quartzite (BMQ):* The iron content of low grade banded magnetite quartzite (BMQ) of Kudremukh and Bababudan regions is about 35 to 38% Fe . This ore cannot be used profitably in its original state. Hence, beneficiation is required to improve its grade to more than 65% Fe and meeting the specifications of pellet plant for smelting/export.
- Pelletisation of the high grade concentrate is a process of agglomeration involving the following steps.
- Finer grinding of the concentrate to attain the requisite fineness;
- Mixing with binders such as bentonite/lime/fluxes and water.
- Balling to produce wet spherical balls measuring about 6 to 16 mm,
- Heat hardening of spherical balls is carried out to produce physically hard pellets meeting the specifications of iron and steel plants. The pellets are used along with the calibrated lumps for iron making in blast furnace/direct reduction furnaces.

Low grade hard hematite (with or without magnetite) ores: Some of the ores that fall under this category include BHQ, BHJ, Taconites, Itaberite, containing about 25 to 35% Fe. The rocks containing such low iron contents are unsuitable for use in their original state. Hence, beneficiation of this low grade ore is essential to improve the concentrate grade to a level of more than 60% Fe and making it suitable for pellet making and smelting the pellets in steel mills.

Reserves of Indian iron ores are reasonably abundant and quite rich but some pockets in the mine may contain undesirable impurities such as high alumina, high phosphorus, and sulphur (> 0.03). Special efforts have to be made, including ore blending, to reduce, not only alumina content, and thus alumina to silica ratio, but also the content of alkalis, phosphorus, and sulphur in the ore. Occasionally, selective mining is carried out to leave out pockets of very high phosphorus and sulphur in mineral/deposits and outcrops.

Typical specifications of iron ore required for sinter making for use in blast furnace:

Al_2O_3	:	2% max (alumina content in Sinter: 2.6%)
Reducibility	:	60-65%
Tumbler index (ISO)	:	75% min
RDI (ISO)	:	25% max

The other ores that require beneficiation to improve their chemical quality include limestone, dolomite, magnesite, apatite, fluorspar, and barytes and so on. These ores require beneficiation not only to improve their chemical quality through reduction of associated gangue minerals like quartz and silicate minerals but also to reduce their size for meeting the specifications/requirements of the ultimate user (iron and steel making). Typical specifications of limestone for use in blast furnace are as follows:

CaO	:	> 50%
MgO	:	0.34%
Al_2O_3	:	0.5%
SiO_2	:	0.2%
LOI	:	44%

Specifications of Dunite preferred by some plants as a substitute to limestone for use in blast furnace are as follows:

MgO	:	49%
SiO_2	:	42%
LOI	:	1 %

Coal (coking and non coking coal)

Indian coals have high ash content and less carbon. Coal washing is therefore undertaken to reduce ash content through separation of lighter clean coal from the heavier gangue minerals (slate, shale, silicates, pyrites, etc.) through use of methods such as heavy media separation, gravity concentration, or flotation.

Diamonds

Kimberlite pipes of Panna (Madhya Pradesh), contain about 10 carats of diamonds (1 carat = 0.2 grams) in 100 tonnes of rock (2 in 10^8 concentration). The pipe rock is crushed in a jaw crusher followed by cone crusher later to achieve its liberation size of -20 mm by working in close circuit with a screen. The -20 + 1.2 mm product is subjected to heavy media separation using ferrosilicon as media. After recovering ferrosilicon by magnetic separation for recycling, the sink product is subjected to X ray sorting. The tailings are further treated with grease tables to recover all the escaped diamonds. The -1.2 mm fraction, the float fraction of HMS, and heavy gangue minerals are rejected.

Nonferrous ores (copper, lead, and zinc ores)

The metal content (or grade) of these ores is around 0.5 to 2%. Ores of these minerals (chalcopyrite and chalcocite) are usually associated with lead ore (galena) or zinc ore (sphalerite). These nonferrous ores can be recovered profitably by beneficiation involving selective flotation. In the alternative, they can be floated together (bulk flotation) followed by their separation into individual minerals (differential flotation).

Usually, these ores are also associated with precious metals like gold, silver, and other valuable elements like nickel, molybdenum, uranium, as well as unwanted minerals like pyrite. It is necessary to recover these precious metals profitably along with sulphides concentrate by their separation during smelting stage. It is not economical to smelt these low grade ores without prior beneficiation/enrichment. Hence, beneficiation by selective /differential flotation or hydrometallurgy techniques is commercially carried out followed by recovery of precious metals during smelting. The flotation tailings are further treated for recovery of nickel, molybdenum, or uranium minerals. These beneficiation plants are located in Ghatsila (Bihar) and Khetri and Jawar (Rajasthan) and so on.

Barytes

Barytes is processed to meet the specifications of oil-drilling wells. The vast reserves of high grade barytes of Mangampet are processed mostly by dry crushing followed by grinding in Raymond mills. In a few cases, if necessary, gravity concentration techniques are utilized to reject associated gangue minerals.

Bauxite

An important aluminium bearing mineral, bauxite is widely available in India. Bauxite is associated with laterite, clay, and other siliceous minerals. The metal is recovered through hydrometallurgical processes.

Uranium bearing ores

These contain very low concentrations of U_3O_8 (triuranium octoxide, a compound of uranium) ranging from 0.05 to 0.1%. These are available in Jaduguda (Bihar) and Tummalapalli (Seemandhra). This element is recovered by gravity or hydrometallurgy and ion exchange processes.

Gold, silver, and platinum (precious metals)

For commercial exploitation of gold ores economically, the metal content of the ore (tenor/grade) should preferably be at least 1 gram per tonne ore. These high value but low grade precious metals are recovered (occasionally by panning river sands) successfully by a combination of gravity separators, hydrometallurgy or pyro-metallurgy or ion exchange processes.

Chapter 3
Iron Ores

3.1 Very high grade hematite ore

High grade hematite ore available in the country, vary between 65 and 69% Fe and has 1 to 3% SiO_2, 1 to 5% Al_2O_3, besides P and S varying between 0.02 to 0.04%. Practically, the entire mine production of hematite ore is suitable for sale/for iron and steelmaking after crushing, sizing, and classification. There is less than 10% mine waste. The -100 +10 mm lumps are mostly exported. The calibrated lumps (-30 +6 mm) are used for smelting by using direct reduction (DR) or blast furnaces (BF) route. The -10 mm + 100/200 mesh fines are either exported as it is or sintered prior to smelting in indigenous blast furnaces. The -200 mesh fines are pelletized prior to smelting for domestic consumption/export market. With improved processing techniques and greater market demand, even these tailings/rejects can be sold in domestic as well as export markets after further enrichment or through blending them with the current high grade fines.

Hematite ores of Bailadila region contain 14 deposits of iron ore, with more than 66% Fe and very low silica and alumina contents. A view of the Bailadila iron ore processing plant operated by National Mineral Development Corporation Limited is shown in Photo 3.1. Processing involves just two or three stages of crushing and sizing to meet the domestic market specifications, and reducing the final rejects/losses to less than 10%. The operation is designed to produce either -200 +10 mm or -100 +10 mm, lumps for export and -30 +6 mm size lumps for domestic market; -10 mm or, -6 mm +200 mesh sinter fines; and -200 mesh iron ore concentrate suitable for pellet making. Each of these mines is designed to produce around 4 to 6 MTPY (million tonnes ROM/year) with more than 90% saleable products. These mines are designed to have the built in scope for expansion. Out of 170 million tonnes of iron ore produced in India during the year

2011, about 20 million tonnes of high grade ore was produced out of this region by NMDC Limited, (an organization owned by the Government of India), for supplying iron ore mostly to the domestic market. Photo 3.2 shows the mining and ore processing sections of Bailadila 5 & 11c at Bacheli.

India is one of the leading producers of high grade iron ores in the world. Due to its abundant reserves, in addition to meeting internal requirements, India has been exporting iron ore on a massive scale/ without any foresight, mostly to Japan, China, South Korea, and other European countries for earning foreign exchange. The unrestricted export both legally and illegally over the past decade has depleted country's reserves of high grade iron ores. Realizing the urgent need for conservation of high grade ores for the use of future generations, Indian Government has imposed a ban on exporting + 63 % Fe high grade ores and hiked the export duty of the lower grade ores.

India's iron ore reserves as on 1.4.2010 are shown as follows:

Mineral	**Reserves (million tonnes)**	**Total production (MTPY)**	**Principal producers**
Hematite	17,882	207.998 (say 208)	NMDC Ltd. (Bailadila 14, 5 &11 C, Donimalai) Sesa Goa Ltd. (Gurmel) Sarda mine (Thakurani) Joda East (Tata Steel Ltd.)
Magnetite	10,644		
Total iron ore	28,526		

Source gratefully acknowledged: Statistical Profiles of Minerals 2010-2011 by Indian Bureau of Mines, Nagpur, dated April 2012.

Processing of very high grade hematite ores of Bailadila deposit is shown in Figure 3.1.

Operating parameters

The ore is of very high grade (+65% Fe) with silica (2-3%), alumina (< 1 %); SiO_2:Al_2O_3 ratio > 1.

Total water consumption: 0.6 to 0.7 M^3 to 1 M^3 water/tonne ROM

Power consumption = about 1 KWH/tonne ROM

Crushing to –100/40/30 mm size using 2/3/4 stages of crushing

Results of processing:

Product	**% Fe**	**Use**
- 40 +10/-30 + 6 mm lumps	65 – 66	BF/DR grade lumps
- 10/-6 mm +100/200 mesh fines	64 – 66	Sinter fines
- 200 + 325/400 mesh cyclone underflow	65 – 68	Pellet fines
Cyclone overflow	< 50	Tailings/rejects
ROM	64 – 66	

Photo 3.1: A view of Bailadila iron ore screening plant

Courtesy: NMDC Limited

Photo 3.2: Crushing, screening, and conveyors at Bailadila 10/11A, Bacheli

Courtesy: NMDC Limited

Figure 3.1: Processing very high grade hematite ores (ores of Bailadila region)

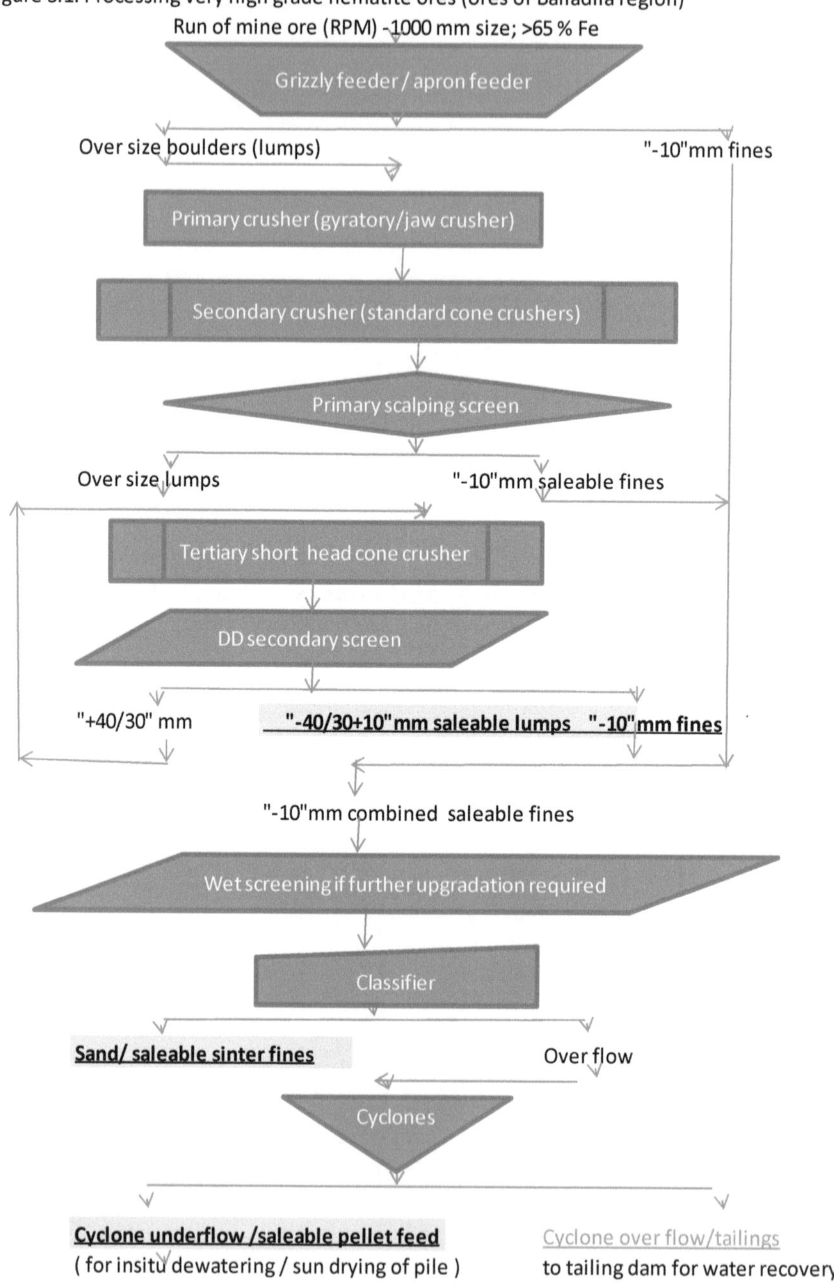

3.2 High grade hematite ores with high alumina content

Iron ore of Kiriburu, Meghataburu mines, captive mines of SAIL, Noamundi mine of Tata Steel, etc., is associated with clay, laterite, and other silicate gangue. Hence, clay adhering to the iron ore is removed by scrubbing or log washing prior to further treatment meant for high grade ores.

Need for scrubbing, classification, and cycloning: The processing of Kiriburu ore involves scrubbing, classification, and cyclones as shown in Figure 3.2. Scrubbing action is essential to remove clay/laterite associated with the iron ore by allowing attrition or rubbing of one lump against the other in the medium of water. This helps in preferential reduction of alumina content along with silica, thereby reducing the gangue content of the saleable iron ore. This also helps with improving the silica to alumina ratio and marginal enhancement of iron content in the ore.

The -30 +10 mm washed lump ore is used as blast furnace feed for iron making.

The -10 mm +100 mesh classifier sand is suitable for use as sinter feed for iron making.

Cycloning of -100 mesh classifier overflow is carried out to recover additional iron values in the form of cyclone underflow for use as pellet feed (after dewatering) for iron making and reducing the iron losses leading to pollution control.

Reduction of alumina content in the iron ore is necessary and desirable for iron making in order to increase blast furnace productivity and reduce coke and flux consumption.

Order of magnitude of results:

Product	**% Fe**	**Usage**
-30 + 6 mm lumps	60 – 62	Blast furnace feed/feed for DR plants
-6 mm + 100 mesh classifier sand	59 – 62	Sinter feed
-100 mesh + 325 mesh cyclone underflow	62 – 64	pellet feed
Cyclone overflow	<50 % Fe	Tailings/rejects
ROM/feed	55 – 59%	

Water requirement: Total 1 M^3 water/tonne ROM (up to 0.3 M^3/tonne water during scrubbing and 0.7 M^3/tonne water for wet screening stages); even more water is required if the clay content is more or the ore has undesirable yellow ochre.

Figure 3.2: Processing of high grade hematite ores with high alumina

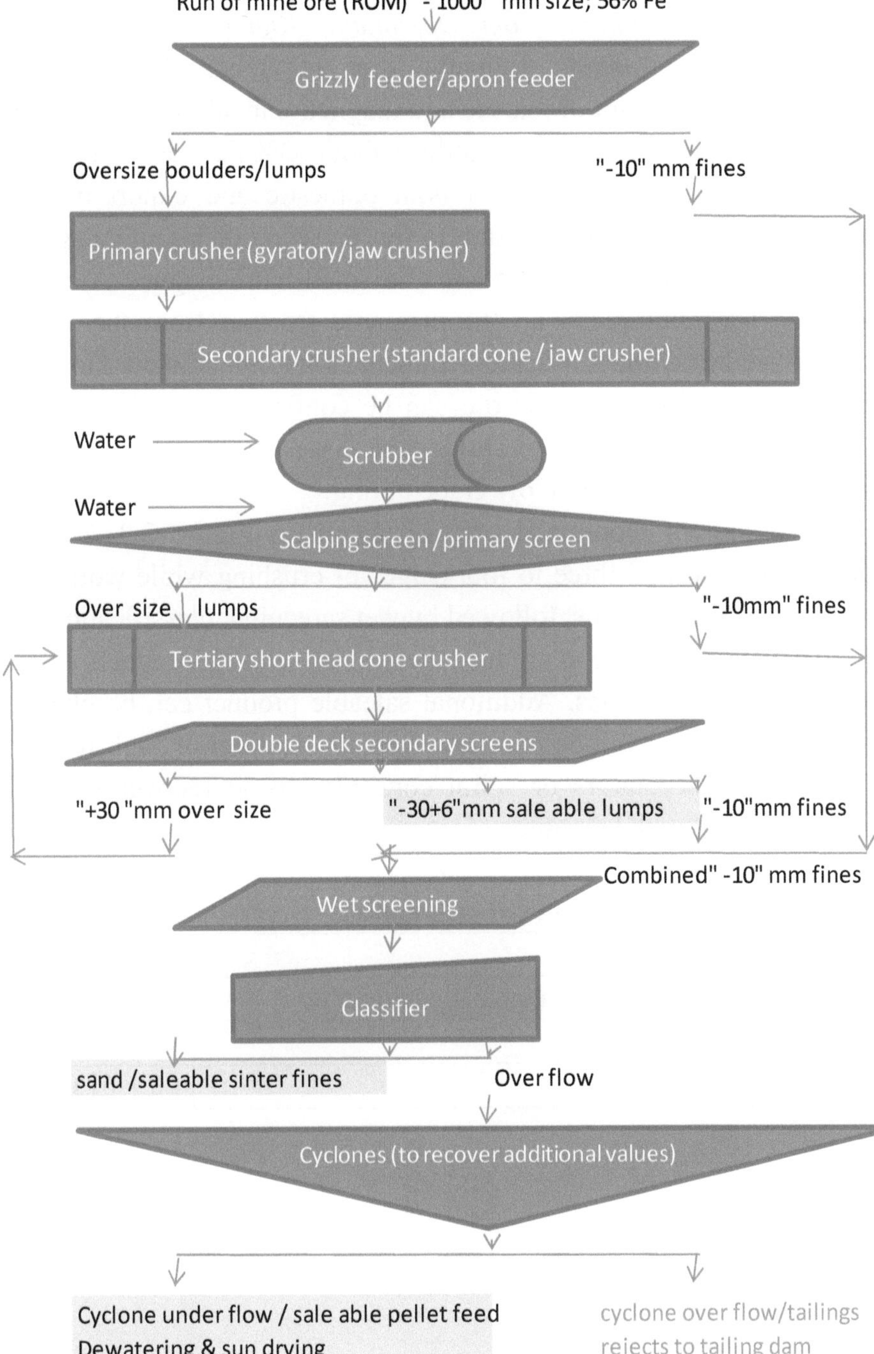

3.3 Processing of medium grade hematite ores

Hematite ores available from Donimalai, Kumaraswamy, Bellary, Hospet, and Goa mines. Hematite ores of Bellary–Hospet region contain a small amount of martite/and magnetite in addition to silicates. They are very widely mined to produce about 63% Fe grade concentrate that meets the specifications of both domestic and export markets. Magnetic property of the ore due to the presence of martite/magnetite can be observed even with the use of a small hand magnet. A cobbler magnetic separator picks up the magnetic fraction from the -6 mm crushed ore but without much enrichment of its iron content. However, the magnetic property of the ore can be confirmed by subjecting the -200 mesh ground sample to Davis tube test at 5000 gauss producing high grade concentrate and low grade tailing.

Processing of Donimalai ore, as shown in Figure 3.3, involves subjecting the ore to three to four stages of crushing while working in close circuit with screens, followed by wet screening, classification, and hydro cyclones in order to produce the desired saleable products such as lumps, fines, and slimes. Additional saleable product can be obtained by processing the cyclone overflow through medium or high intensity wet magnetic separators or spiral concentrators to recover magnetic particles locked with hematite.

Results of beneficiation (order of magnitude):

Product	% Fe
Combined hematite and magnetic concentrate	+63
Rejects	<20
Head	60 – 61.0

Figure 3.3: Processing medium grade hematite ores

(Ex. Medium grade hematite ore of Bellary-Hospet region associated with magnetic minerals)

ROM (Soft/ medium grade hematite ore)

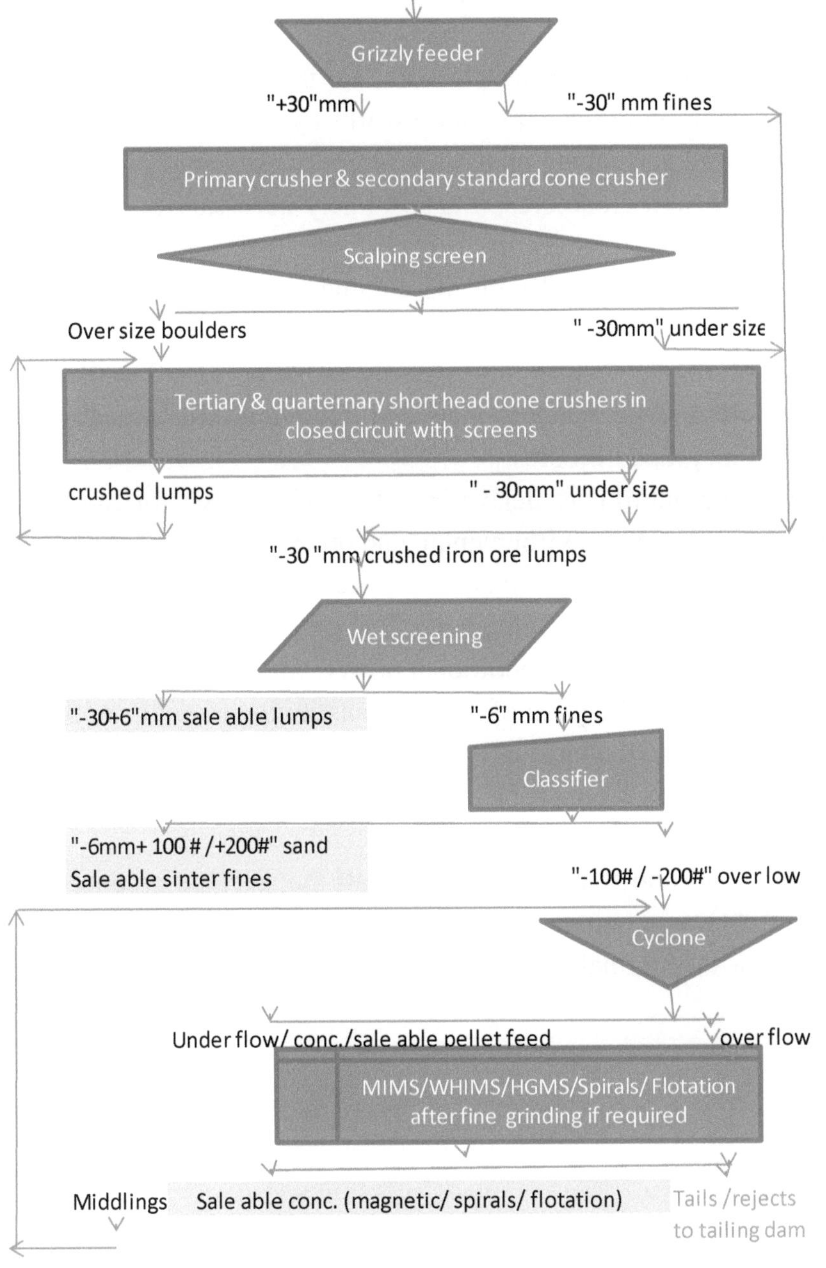

3.4.1 Banded hematite quartzite (BHQ)

Due to extensive exploitation of high grade hematite ore, the available reserves of high grade iron ores are drastically reduced and perhaps may just last for the next 30 years or so. The alternative solution is undertaking mining to a greater depths for tapping additional resources, if available or on-going exploration works by several mining companies resulting with substantial additional reserves. As a part of conservation of available high grade ores, it is absolutely necessary to make the best use of the low grade hematite ore available abundantly in the country. The problems associated with the commercial exploitation of the low grade iron ores are as follows:

Identifying the techno-economically viable process flowsheet.

Very high capital investments and operating costs as compared to normal high grade iron ores.

Difficulty/inordinate delays in getting clearances for mining from Government, forest, environmental, and other statutory agencies.

Acquisition of land outside reserve forest areas containing the mineral deposits, for ore processing.

Getting permission of additional power, water, and tailings disposal.

Commercial utilization of tailings constituting more than 50% of the low grade feed to the plant.

Ore transportation problems between mine to process plant and process plant to pelletization plants/concentrate buyers.

Less profitability as the final yield of the saleable product is much less (<50%) as compared to the massive ore with less than 10% loss coupled with minimal capital investment and less operating costs and utilities.

Order of magnitude of results with reverse cationic flotation of silicate gangue out of BHQ ore

Product	% Fe	Reagents used
Cleaner Hematite Concentrate	>58 – 60	Cationic flotation; very fine grinding to -325 mesh; Amine: 0.1 to 0.2 kg/T; Starch: 0.5 to 0.8 kg/T; pH: 9-10.5; 15 to 20% S
Middlings	40 – 42	
Tailings	<16 – 18	
Head	35 – 38	

3.4.2 Banded hematite jaspar (BHJ) of Donimalai mine

As per the available press reports, processing BHJ ores of Donimalai Iron Ore Mine, as shown in Figure 3.4, involves grinding the ore to –150/–200 mesh followed by spiral concentration and high intensity magnetic separation with or without high gradient magnetic separation of the spiral tails. NMDC limited proposes to use the BHJ ores after beneficiation, for the first time in India, along with the existing dump fines producing 0.36 MTPY iron ore concentrate to meet a part of the pellet feed requirements for the new pellet plant of 1.2 MTPY under construction. Besides increasing the quantity of value added product leading to increased profits and country's resources base, actual usage of the hitherto BHJ waste is expected to minimize the pollution problem caused by the existing waste dumps of BHJ and old iron ore fines/slimes.

Figure 3.4: Beneficiation of BHQ / BHJ ores

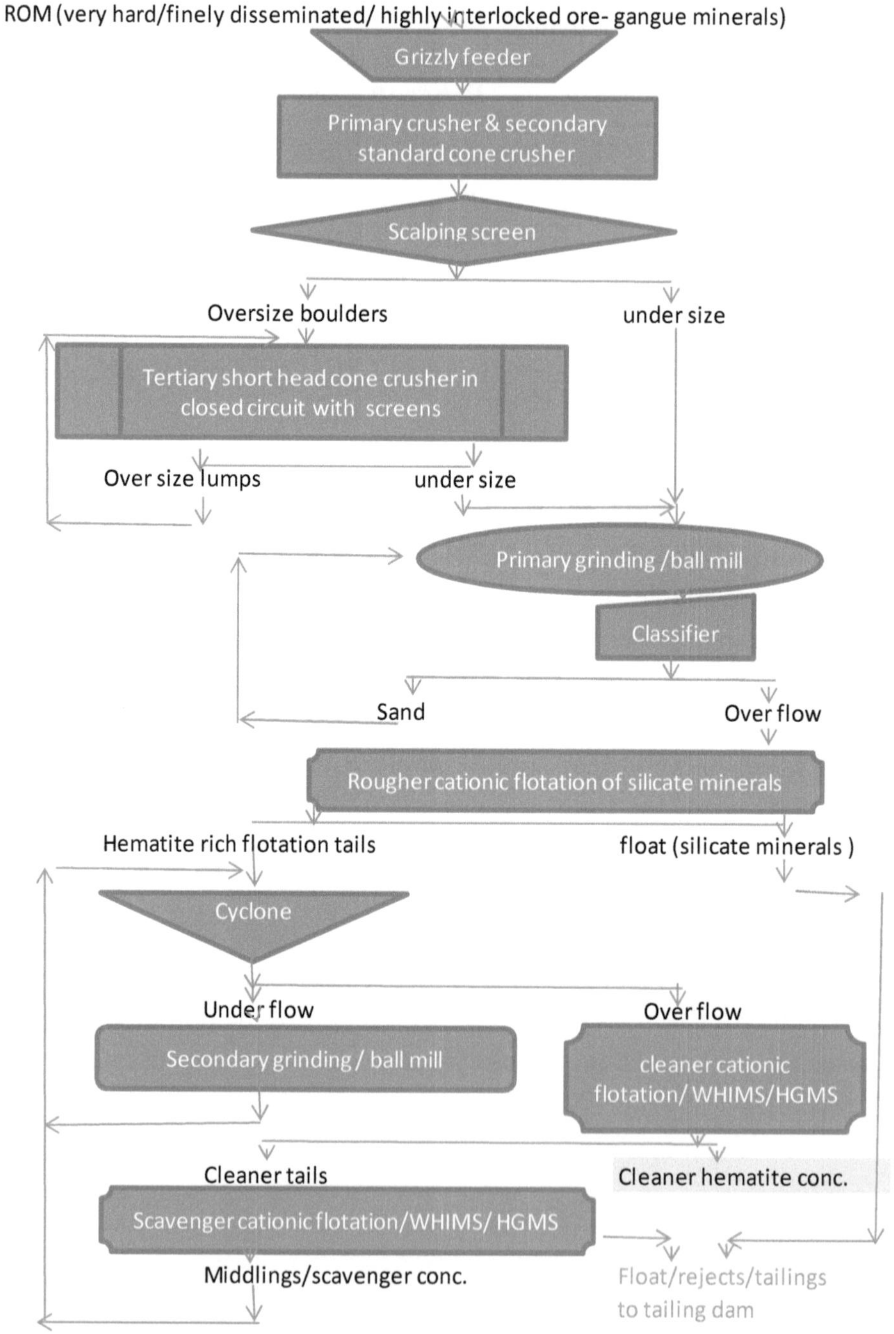

ROM (very hard/finely disseminated/ highly interlocked ore- gangue minerals)

Grizzly feeder

Primary crusher & secondary standard cone crusher

Scalping screen

Oversize boulders

under size

Tertiary short head cone crusher in closed circuit with screens

Over size lumps

under size

Primary grinding /ball mill

Classifier

Sand

Over flow

Rougher cationic flotation of silicate minerals

Hematite rich flotation tails

float (silicate minerals)

Cyclone

Under flow

Over flow

Secondary grinding / ball mill

cleaner cationic flotation/ WHIMS/HGMS

Cleaner tails

Cleaner hematite conc.

Scavenger cationic flotation/WHIMS/ HGMS

Middlings/scavenger conc.

Float/rejects/tailings to tailing dam

3.5 Low grade banded magnetite quartzite (BMQ ores)

KIOCL (Kudremukh Iron Ore Company, Kudremukh, Karnataka) had initially around 610 million tonnes of weathered BMQ ore of about 38.6% Fe and 410 million tonnes of hard fresh ore. The ore contained an admixture of magnetite and hematite. KIOCL exploited the BMQ ore from September 1980 onwards till the end of 2005 when the production was halted by the Supreme Court of India in order to protect environment. The plant was designed to produce 7.5 MTPY iron ore concentrate/pellet feed of + 66.5% Fe grade and less than 2.5% combined silica and alumina, processing more than 20 MTPY ROM with an yield of 38%. The overall project cost at the time of project approval in the year 1975 was US$ 630 million. The initial beneficiation process, as shown in Figure 3.5, involved the following operations.

Crushing, and wet grinding in autogenous grinding mills (with no grinding media) working in close circuit with screens.

Low intensity magnetic separation at 900 to 1000 gauss to recover magnetite.

Rougher, cleaner, and re cleaner stages of spiral concentration for recovery of total hematite along with interlocked magnetite.

Regrinding the rougher concentrate to about 60 to 65% –325 mesh, followed by cleaner and re cleaner magnetic separation to produce desirable high grade concentrate.

Transportation of the final slurry concentrate/pellet feed of +66% Fe grade, with a single-stage pumping through an inclined pipeline over a distance of 75 km from Kudremukh to Mangalore port.

Thickening and filtration of the concentrate slurry at Mangalore port followed by export of the concentrate etc.

With progress of mining to lower depths and change of ROM quality, several modifications, including semi autogenous grinding (with grinding media) of the crushed feed, adoption of finer grinding, replacement of spiral concentrators with cationic flotation, Derrick screens, and replacement of disc filters with horizontal filters etc., were incorporated.

KIOCL has set up a pelletization plant of 3 MTPY capacity for agglomeration of the final filter cake at Mangalore port for meeting export commitments.

Initially, the feed for processing consisted of a combination of weathered BMQ (say 70%) and hard fresh ore (30%). In autogenous mills, the hard primary rock acted as a grinding media to reduce the particle size of the soft, weathered ore particle to its liberation size. In the later years, grinding balls had to be added subsequently to achieve liberation of the ROM containing higher proportion of fresh ore to the weathered ore than what was designed or envisaged initially.

Low intensity magnetic separation at about 900 gauss recovered most of the magnetite particles. To improve its grade further to 65–67% Fe, grinding followed by cleaner magnetic separation was undertaken. Regrinding helped in further liberation of the locked hematite particle associated with magnetite and quartz besides meeting specific surface area requirements (around 1500 sq.cm./gram) of the pellet feed prior to pelletization. Finer grinding also helped the slurry transportation of the final concentrate, without sedimentation, from Kudremukh to thickening and filtration plant/pellet plant located at Mangalore.

The coarse hematite particles (-48 +100 mesh) used to be recovered from the non magnetic tailings by gravity methods of concentration like spiral concentration. As the hematite particles, currently were of very fine size, cationic flotation had to be adopted for recovery. Overall power consumption was 60 KWH/tonne concentrate.

Order of magnitude of results

Product	Weight (%)	% Fe	Separation process/reagents used
Magnetite concentrate	25	67	Magnetic separation
Cleaner hematite Concentrate	15–20	65	Cationic flotation, Grind : 60% -325 mesh Amine: 0.1 to 0.2 kg/T, Starch : 0.5 kg/T, pH: 10 to 10.5
Tailings	55–60	15	For water recovery
Head	100	38.75	Weathered BHQ containing magnetite and hematite

Figure 3.5: Beneficiation of low grade BMQ ores associated with hematite

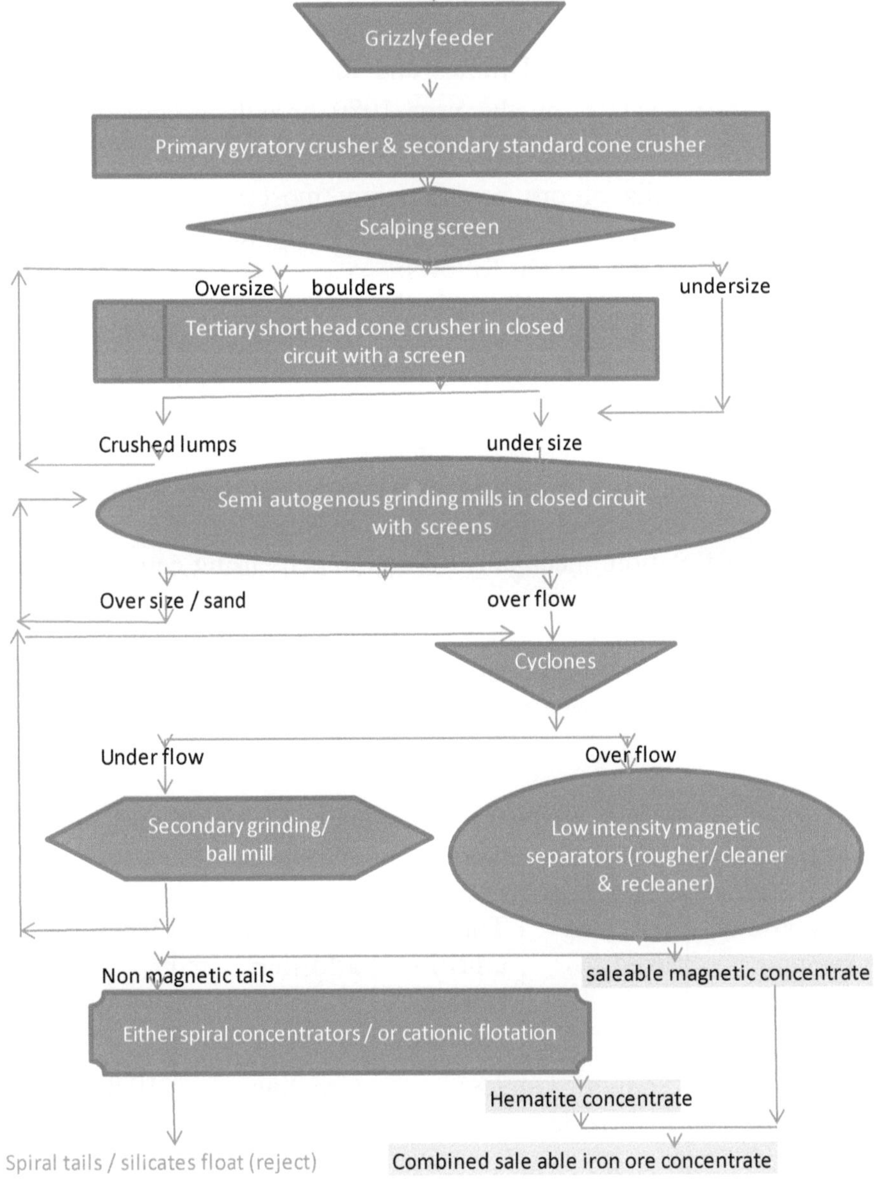

(further grade improvement of the concentrate through Derrick screening at the later stage)

3.6 Beneficiation of low grade and hard hematite/magnetite ores (taconite)

USA has successfully exploited its extensive low grade taconite ores on a massive scale in several mines and increased its pellet production capacity to 57 MTPY by the year 1980 broadly as per the general flowsheet shown in Figure 3.6.

General details of ore beneficiation, most widely adopted on a massive commercial scale in USA, for low grade hematite/magnetite ores by Tilden process are as follows:

Required grind: Very fine grinding: 95% -325 mesh (78%, - 500 mesh).

Process: Selective flocculation of fine iron minerals and dispersion of quartz followed by de-sliming off the silicate minerals in a thickener as overflow.

Reverse cationic flotation of locked silicate minerals, after regrinding the underflow, at around 10.5 to 11 pH.

Type of flotation: mechanical flotation cells using cationic flotation reagents.

Number of stages of rougher/scavenger/cleaner flotation used: 6

Scale of processing/production: 4/8/ 12 MTPY

Order of capital costs: quite high, yet profitable commercially on a massive scale.

Order of magnitude of flotation reagents required and operating parameters:

Purpose of reagent	Reagent used	Total addition Lb/tonne feed	Places of addition	Remarks
pH regulator	Sodium hydroxide	2.85	Grinding, de-sliming, flotation	10.8 –11 (de-sliming at 10. to 10.5 pH)
Flocculent for hematite	Starch	1.75	Flocculation, de-sliming, conditioning	

Depressant for silicate minerals	Sodium silicate/ Glass H	0.25	Ahead of flotation	
Cationic collector for quartz	Amine	0.35	Ahead of flotation	
Flocculent used in thickener for dewatering	Polymer	0.5 –1.0		

Order of magnitude of results of beneficiation

Product	%
Final hematite/magnetite concentrate (flotation tails)	64 – 66 Fe
Slimes	7 Fe
Final flotation silicates concentrate/tails (rejects)	16 – 17 Fe
Combined tailings	11 – 12 Fe
Combined weight percentage of rejects	55 to 65%
Feed grade	20 – 35 Fe

Overall reclaimable water after dewatering/filtration for recirculation: 95%.

Production capacity: 8 MTPY pellets.

Total ore, rock, and overburden to be excavated: 36 million tons per year for 8 MTPY concentrate production.

Figure 3.6: Beneficiation of low grade & hard hematite/magnetite ores (taconites)

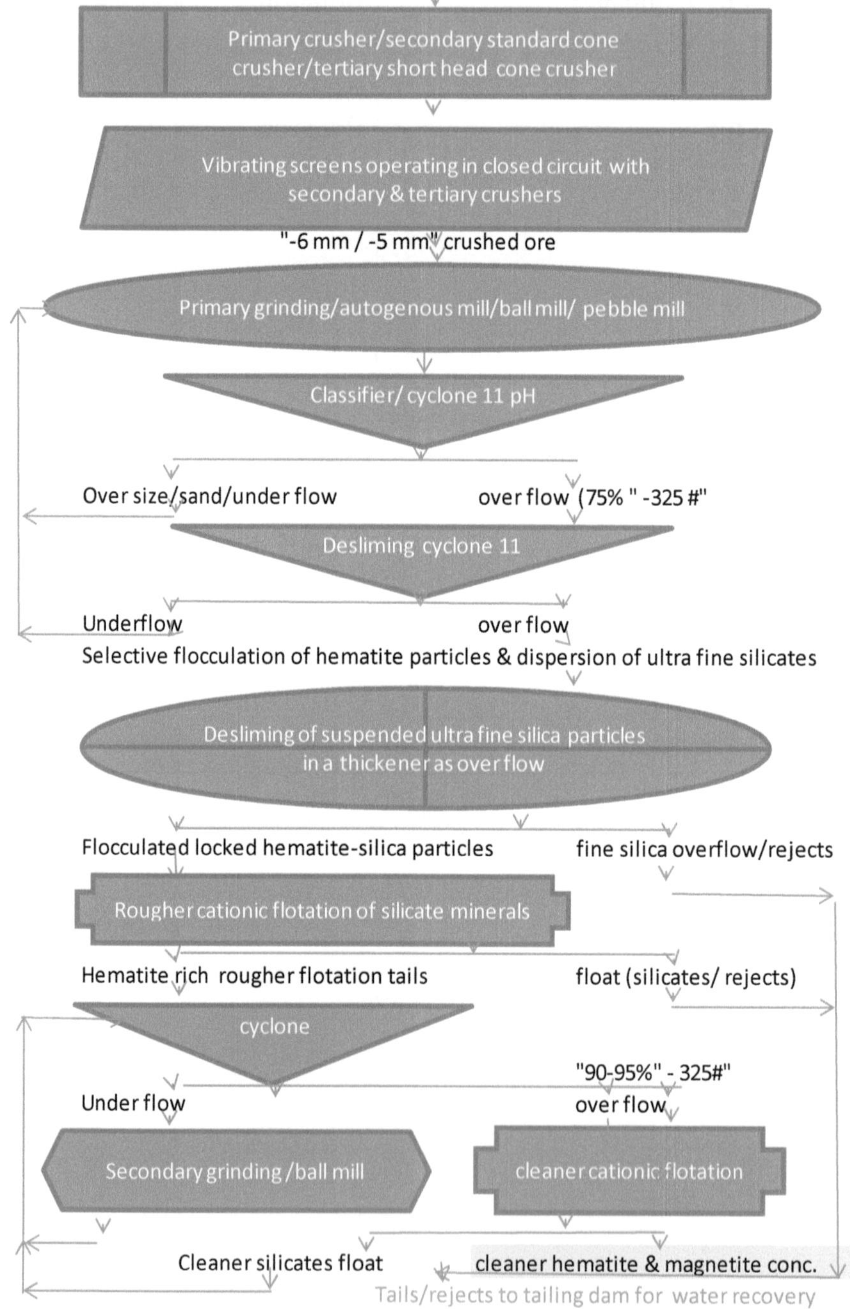

3.7 Low grade hematite ores (Itaberites of Brazil)

Ore reserves: Several billion tonnes

Grade: 44- 45% Fe

Operations include mining on a massive scale, beneficiation, slurry transportation covering a distance of more than 400 km, pelletization, and shipment of pellets.

Beneficiation process. It involves crushing, screening, and classification. The classifier overflow is subjected to grinding, working in close circuit with hydro cyclones, to produce an overflow for liberating iron bearing minerals from the associated gangue. The overflow is subjected to reverse conventional/mechanical flotation producing a rougher concentrate of iron bearing minerals. The rougher concentrate is reground in close circuit with secondary cyclones for achieving complete liberation of iron bearing minerals from the associated gangue. The secondary cyclone overflow is cleaned by reverse column flotation producing final concentrate/pellet feed. Thus, the beneficiation process involves the following major operations.

- Crushing the ore boulders to about -2.5 mm size in multiple stages of crushing, working in close circuit with sizing equipment.
- Closed circuit wet grinding of -2.5 mm crushed ore to 80% -325 mesh
- Rougher cationic flotation in conventional mechanical flotation cells
- Regrinding the rougher concentrate to the complete liberation size
- Cleaning the reground concentrate by reverse column flotation at about 10.5 pH.

The plant recovers about 57 to 59% iron present in the feed.

The plant has huge capacity to produce 20 to 32 MTPY iron ore concentrate/pellet feed from a feed of about 44-46% Fe.

The concentrate is dewatered, pelletized, and exported.

A general flowchart for the beneficiation of low grade Itaberites of Brazil is shown in Figure 3.7.

Order of magnitude results

Product	Weight (%)	% Fe
Iron ore concentrate/pellet feed/flotation tailing	58	67
Middling/silica float/flotation concentrate/rejects	27	20
Tailings/slimes/rejects	15	12
ROM	100	46.06

Figure 3.7: Beneficiation of low grade hematite ore (Itaberites)

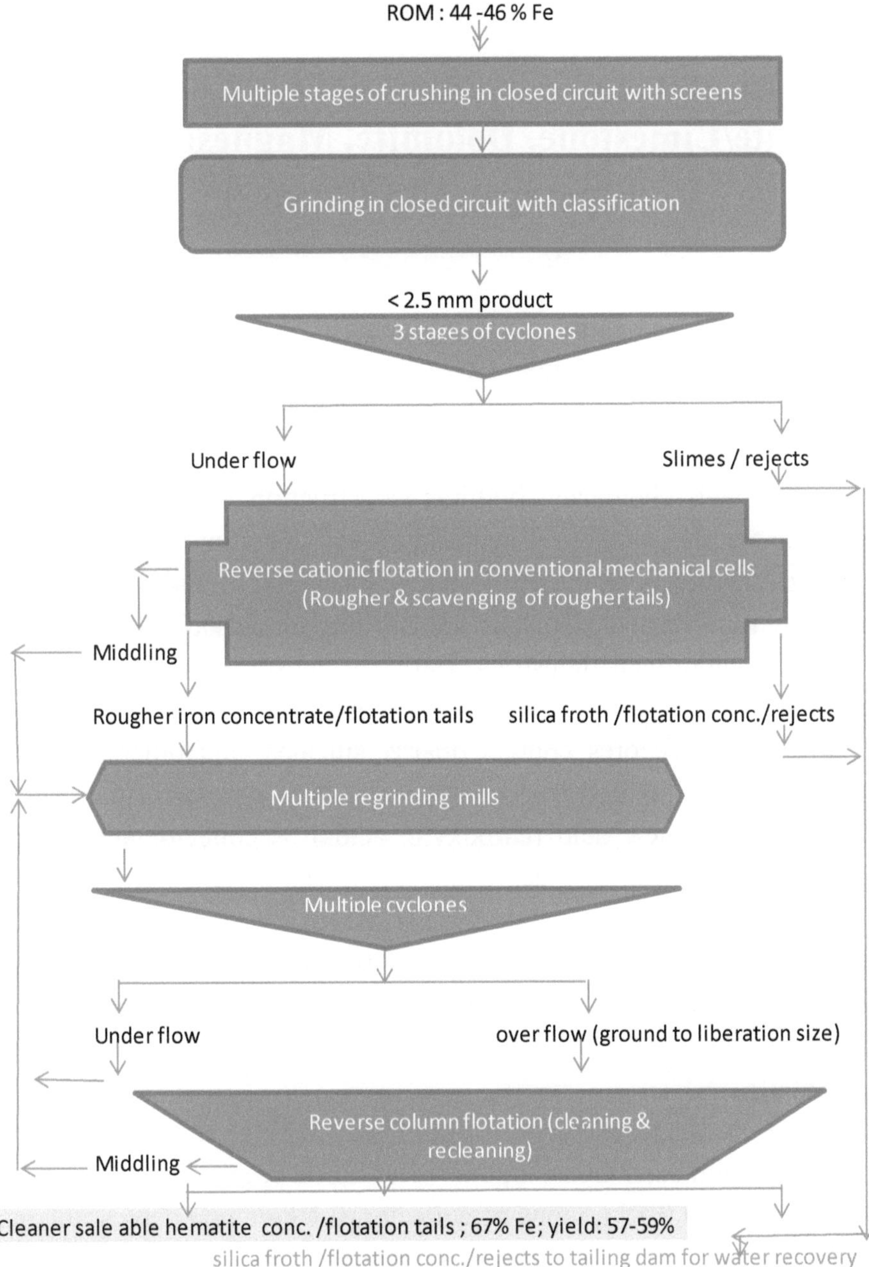

Source of information: Internet - gratefully acknowledged

Chapter 4

Ores of Alkaline-Earth Metals (Calcite/Limestone, Dolomite, Magnesite, Apatite, Fluorspar and Barytes)

Ores of alkaline-earth metals include the following: Calcite ($CaCO_3$)/limestone, dolomite ($CaCO_3MgCO_3$), magnesite ($MgCO_3$), apatite (Ca_4) (CaF) $(PO_4)_3$, fluorite (CaF_2)/fluorspar, and barytes ($BaSO_4$). Calcite is used as a source for making cement, flux in making iron and steel, chemicals, building construction etc. Magnesite and dolomite are the sources of magnesium and are used as flux for steel making. Apatite is a source of phosphorus and is used in fertilizer industry. Fluorite/and fluorspar are the important sources of fluorine used in metallurgical industries. Baryte is mostly used in oil drilling industries.

Most of these ores contain quartz, silicates, and other carbonate minerals as well as hydrated iron oxides. They are easily amenable to flotation using oleic acid (carboxylic acids) as collector and sodium silicate or quebracho as a depressant of silicate minerals in an alkaline medium using lime to regulate the pH. Normally, oleic acid being not so pure acts not only as a collector but also as a frother. As an alternative, amines can also be considered for use as collectors.

4.1 Calcite

For flotation using oleic acid, the ore needs to be ground to about -100 mesh/liberation size. Consumption of flotation reagents is in the order of about 0.25 to 0.5 kg/tonne of collector, 0.1 kg/tonne depressant, and 0.5 kg/tonne pH regulator. The rougher concentrate needs to be cleaned once or twice to meet the flux specifications of the blast furnace or open hearth steel operators. The concentrate containing low silica (<2%) and high CaO (>48%) has to be dewatered, filtered, dried, and agglomerated (either by briquetting or pelletization) before its end use. Limestone beneficiation is carried out as per the general flowsheet shown in Figure 4.1. TISCO's plant at Birmitrapur and SAIL's plant at Nandini generally follow this process route.

As Indian fluxes are of poor quality with high gangue/high LOI, Indian steel plants use around 10 kg/tonne each of limestone and dolomite. Some plants (TISCO) prefer dunite in place of dolomite.

NMDC Limited, a Government of India organization, is proposing to exploit the low silica limestone available in Arki deposit, Himachal Pradesh. However, regular operations have not yet commenced pending statutory clearances from the government agencies and court decisions in favour of starting production and lack of adequate infrastructure such as facilities to transport concentrate by road/conveyor belts.

Total reserves of calcite and limestone

Mineral	**Reserves (million tonnes)**	**Production (million tonnes per annum)**	**Three main producers**
Calcite as on 1st April 2010	20.945	0.03937	Rabachhi, Udaipur (Rajasthan), Sachli, Pindwara, Lohagarh, Perwa (all located in Rajasthan)
Limestone as on 1st April 2005	175345	237.774	Mostly from Nimbettiand, Amli from Rajasthan; Thummalapenta (Seemandhra); Injepalli & Adityanagar (Karnataka)
Combined	175365.945	237.813	

Source gratefully acknowledged: Statistical Profiles of Minerals 2010–2011 by Indian Bureau of Mines, Nagpur, April 2012.

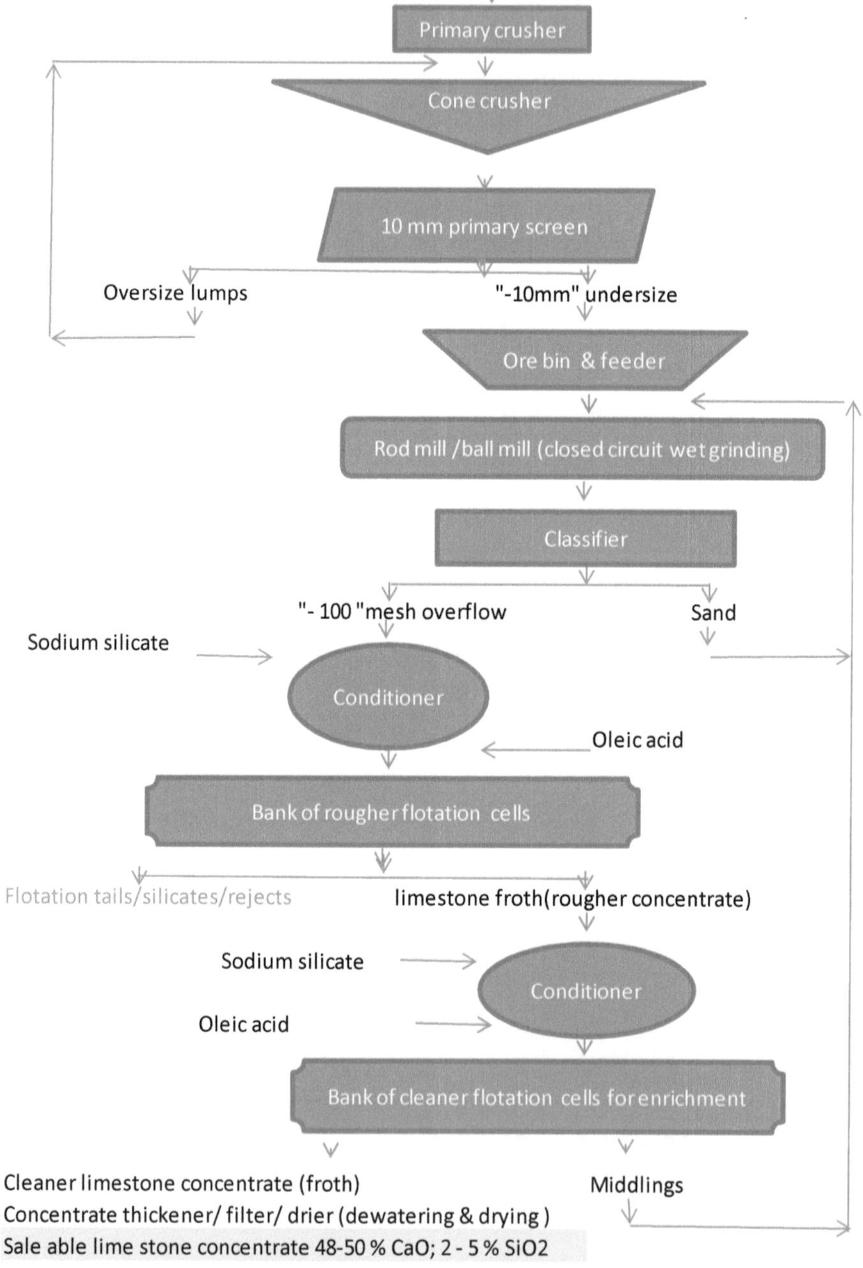
Figure 4.1: Beneficiation of low grade calcite / limestone
ROM (mixture of limestone with quartz/silicate minerals), fairly coarse liberation
Primary crusher
Cone crusher
10 mm primary screen
Oversize lumps
"-10mm" undersize
Ore bin & feeder
Rod mill /ball mill (closed circuit wet grinding)
Classifier
"- 100 "mesh overflow
Sand
Sodium silicate
Conditioner
Oleic acid
Bank of rougher flotation cells
Flotation tails/silicates/rejects
limestone froth(rougher concentrate)
Sodium silicate
Conditioner
Oleic acid
Bank of cleaner flotation cells for enrichment
Cleaner limestone concentrate (froth)
Middlings
Concentrate thickener/ filter/ drier (dewatering & drying)
Sale able lime stone concentrate 48-50 % CaO; 2 - 5 % SiO2

4.2 Magnesite/dolomite

Magnesite and dolomite are the important sources of magnesium. Magnesium chloride present in sea water is also another main source of magnesium. Magnesium has adequate workability characteristics for it to be used as an important structural element for building aeroplanes and in alloys etc.

Impurities contained in magnesite/dolomite include talc, mica, chlorite, and, sometimes, quartz. Since talc is a naturally floating mineral that does not require a collector for its removal, it is desirable to float off talc along with other silicate minerals from magnesite by first using amine as a collector and a frother. Another option is to float magnesite and dolomite preferably from calcite using oleic acid (carboxylic agents) at a high pH of 10, using lime and sodium silicate to depress silicate gangue. To cut down process costs, cheaper collectors like fish oil, corn oil, and tall oils can also be used for flotation though they are not as effective as oleic acid. Furthermore, if oleic acid is used for flotation, frother is not required. Flotation is effective in alkaline medium combined with caustic soda and sodium metasilicate. Other depressants like sodium carbonate, sodium sulphide, polyphosphates, and quebracho can also help in preventing the flotation of siliceous, aluminous, and ferruginous slimes. There is a disadvantage in this method, however, because floating of carbonate minerals available in the ore requires more reagents to depress even small quantities of silicates, which may increase operational costs.

In addition to process of floating talc by first using a frother, another option is to float talc along with silicates with amines such as laurylamine. Impurities can be floated with amines, while the desired mineral, magnesite, can be separated as nonfloat material. This is useful to upgrade magnesite and apatite associated with silicates.

In case magnesite and dolomite are to be separated from calcite, selective flotation can be attempted by using oleic acid, at pH 10, using lime as a pH regulator, and sodium silicate to depress calcite as per the process illustrated in Figure 4.2.

Reserves of magnesite and dolomite as on 1st April 2010

Mineral	Reserves (million tonnes)	Production (million tonnes per annum)	Three main producers
Magnesite	335.172	0.23	Salem (Tamil Nadu), Jhiroli (Uttarakhand),Chettichavadi (Tamil Nadu)
Dolomite	7731	5.065	Bilaspur (Chhattisgarh), Birmitrapur Sundergarh (Odisha); Madharam, Khammam (Telangana)

Source gratefully acknowledged: Statistical Profiles of Minerals 2010-2011 by Indian Bureau of Mines, Nagpur, April 2012.

Figure 4.2: Beneficiation of magnesite/dolomite ore

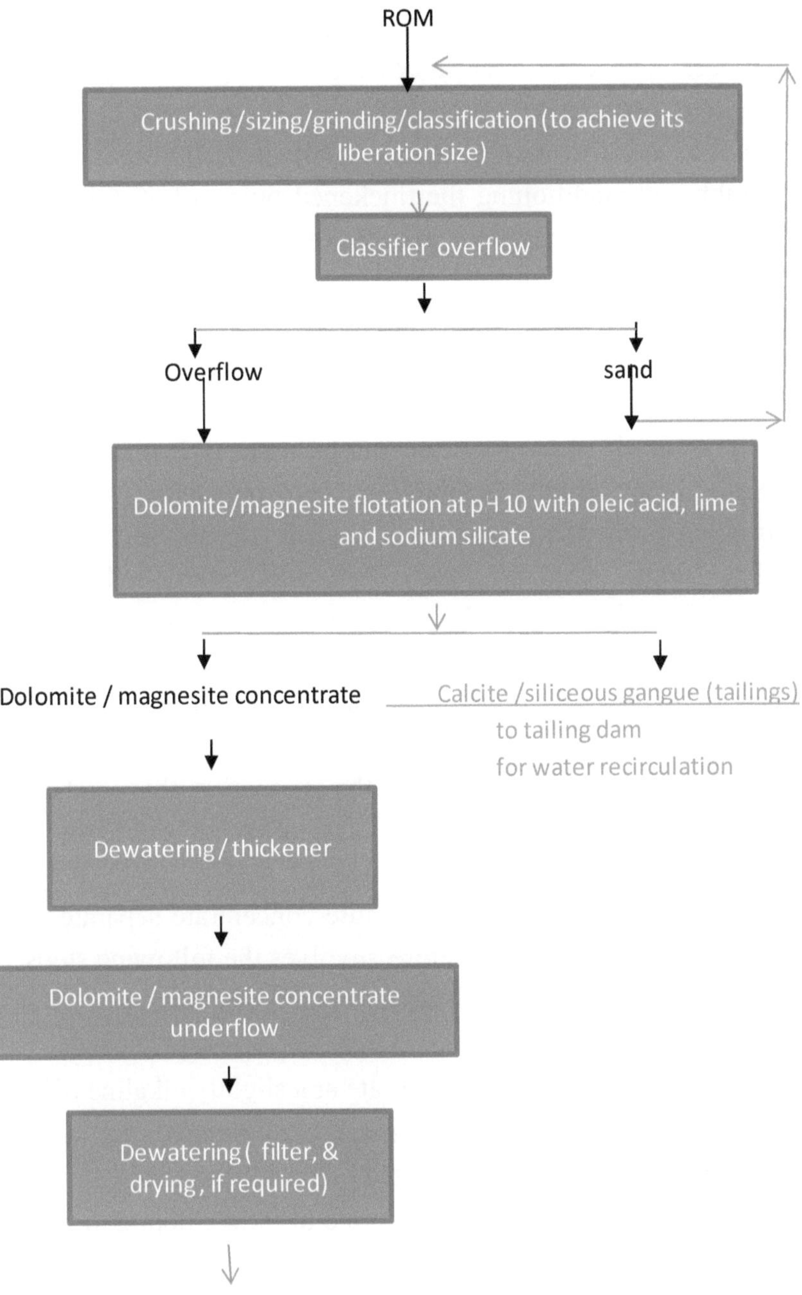

4.3 Apatite

Apatite (calcium phosphate), the source of phosphate for making fertilizers, is associated with quartz, silicates, carbonates, and hydrated iron oxides. Its demand is increasing with growth in population requiring to boost agricultural production.

Generally, beneficiation of apatite involves desliming the crude ore after scrubbing, conditioning the thickened pulp (40-50% solids) with the flotation reagents, and undertaking flotation at a lower pulp density (i.e., after dilution with water). Apatite is beneficiated by flotation as per the general flowsheet shown in Figure 4.3, using oleic acid (or other carboxylic acids such as fish oil, corn oil, and tall oil) in a slightly alkaline medium such as caustic soda. To depress ferruginous gangue minerals, tannin or sodium silicate can be used. A frother may not be required, as commercially available cheaper but impure collectors do have frothing properties. Otherwise, fuel oil can be used. Flotation of crystalline apatite can be carried out with soap, sodium silicate (depressant for silicate minerals), and sodium carbonate (pH modifier).

Another option is silica can be floated by cationic collectors at moderate pH. However, flotation of quartz while depressing apatite is not adopted commercially, as it is not desirable to float the major constituent of the mineral mixture. In the alternative, this can be carried out for cleaning the rougher apatite concentrate produced by anionic flotation, followed by cationic flotation of quartz, which helps with producing a higher grade saleable apatite concentrate separated as non float material. The process, therefore, involves the following steps:

- Conditioning the thicker pulp with flotation reagents.
- Anionic flotation of apatite while depressing silicates, which produce a rougher apatite concentrate at a slightly alkaline pH.
- Acid treatment of rougher concentrate to remove the coating of collector, carboxylic, and associated oils from the apatite.
- Cationic flotation of quartz at moderate pH while depressing apatite.

Ore reserves of apatite as on 1st April 2010

Mineral	Reserves (million tonnes)	Production (million tonnes per annum)	Grade: P_2O_5	Main producers
Apatite	24.229	0.003845	16-20	N.R. Puram, Visakhapatnam (Seemandhra); Beldih, Purulia District (West Bengal); by West Bengal Mineral Development and Trading Corporation Ltd., Kolkata

Source gratefully acknowledged: Statistical Profiles of Minerals 2010-2011 by Indian Bureau of Mines, Nagpur, April 2012.

Figure 4.3: Beneficiation of low grade apatite

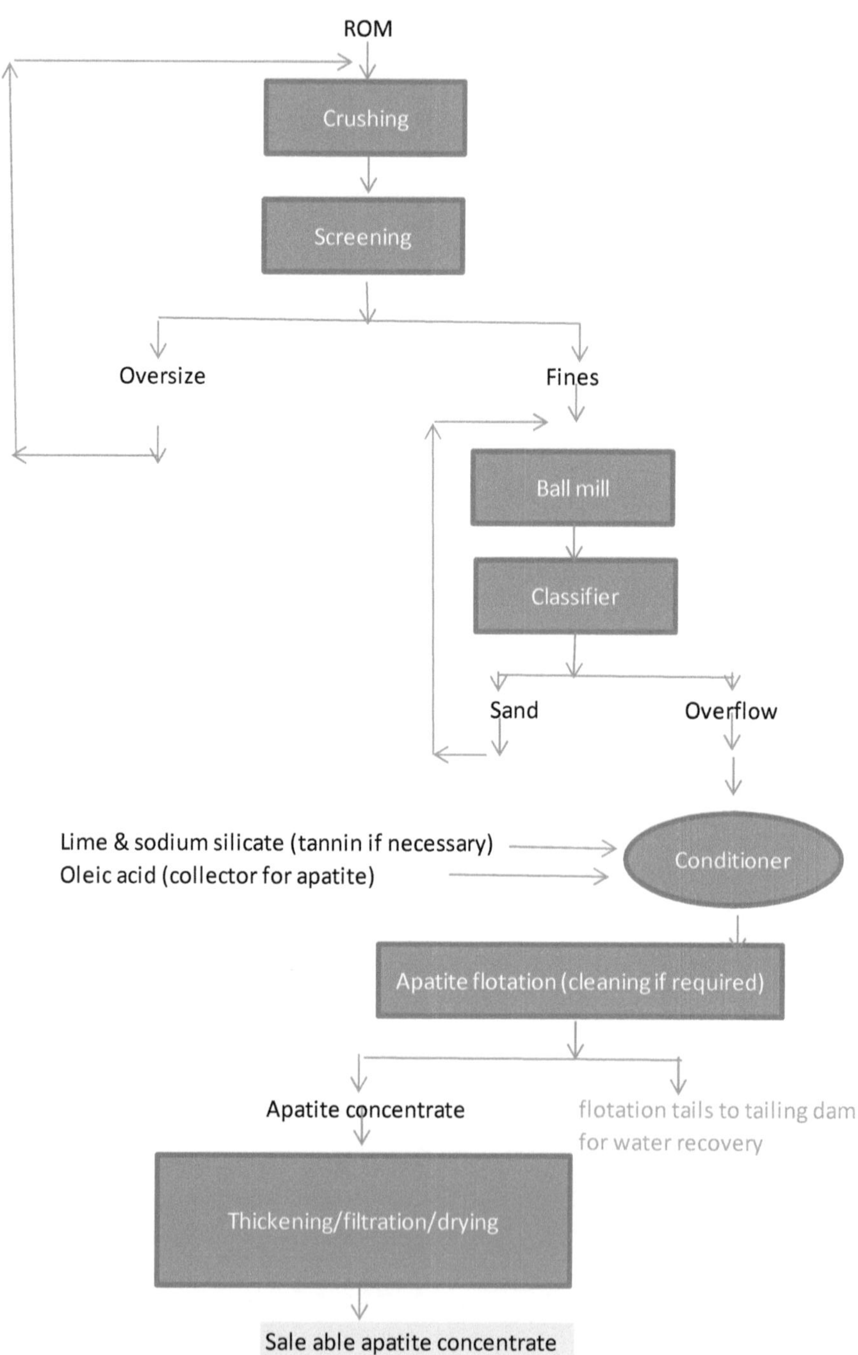

4.4 Fluorspar

Besides fluorite (CaF_2), fluorspar ore is associated with quartz, calcite, barytes, and sulphide minerals like pyrite and sphalerite. Acid grade fluorspar requires very high purity (>96% CaF_2) whereas lower grade coarser concentrate is acceptable for the metallurgical industries (>88% CaF_2).

Beneficiation technique. This involves upgradation of fluorspar by flotation using oleic acid (or other carboxylic acids), as shown in Figure 4.4., preferably at an elevated temperature and neutral pH followed by five to six cleaning stages, using soft water. If the ore is associated with sulphide minerals, it is desirable to float sulphide minerals with xanthates first, prior to flotation of fluorspar by using oleic acid. In the alternative, bulk flotation of fluorspar and sulphide minerals together can be undertaken first, followed by differential flotation of fluorspar using depressants like sodium sulphide for sulphide minerals specifically. Tannin can be used as a depressant for calcite and barytes while floating fluorspar with oleic acid. Although fluorspar floats easily, removal of associated minerals to produce acid grade product is quite difficult/and cumbersome.

Ore reserves of fluorspar as on 1st April 2010

Mineral	Reserves (million tonnes)	Production (million tonnes per annum)	Grade	Main producers
Fluorite	18.214	0.007544 (Graded:0.003150; Concentrate:0.004394) 3.5% acid grade concentrate and 96.5% metallurgical grade	>96% CaF_2 >88% CaF_2	Ambadungar (Gujarat) by GMDC Ltd., for concentrate; Dungargaon (Maharashtra) and Karara (Rajasthan); for graded

Source gratefully acknowledged: Statistical Profiles of Minerals 2010-2011 by Indian Bureau of Mines, Nagpur, April 2012.

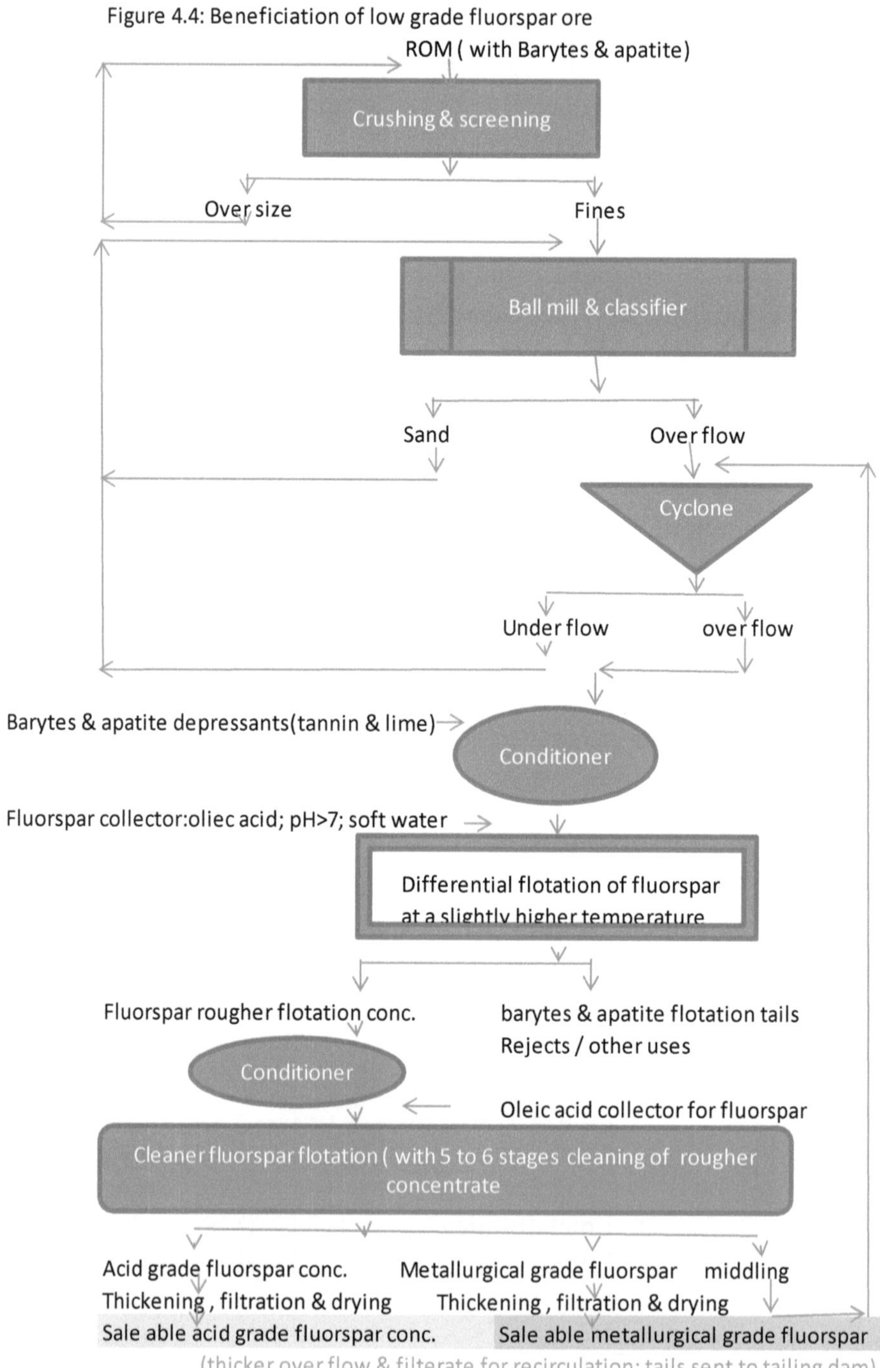
Figure 4.4: Beneficiation of low grade fluorspar ore
ROM (with Barytes & apatite)
Crushing & screening
Over size
Fines
Ball mill & classifier
Sand
Over flow
Cyclone
Under flow
over flow
Barytes & apatite depressants(tannin & lime)
Conditioner
Fluorspar collector:oliec acid; pH>7; soft water
Differential flotation of fluorspar at a slightly higher temperature
Fluorspar rougher flotation conc.
barytes & apatite flotation tails
Rejects / other uses
Conditioner
Oleic acid collector for fluorspar
Cleaner fluorspar flotation (with 5 to 6 stages cleaning of rougher concentrate
Acid grade fluorspar conc.
Metallurgical grade fluorspar
middling
Thickening , filtration & drying
Thickening , filtration & drying
Sale able acid grade fluorspar conc.
Sale able metallurgical grade fluorspar
(thicker over flow & filterate for recirculation; tails sent to tailing dam)

4.5 Barytes

Barytes is usually associated with silicates, iron oxides, fluorite, and calcite.

In Cuddapah district of Andhra Pradesh, several dry processing units, including Raymond mills, with/without air classifiers are in operation for meeting oil drilling requirements of both domestic and export markets. Barytes is associated with quartz, but it is softer and heavier than quartz. Large lumps of 8” size ore are crushed in jaw crushers to about 2” size, which are further crushed in roll crushers to reduce their size to –6 mm. By dry grinding the -6 mm ore in Raymond mill (followed by pneumatic classification in a few cases, if required), finely ground heavier barytes can easily be separated from the coarse and lighter quartz using air classification. If barytes is pure, grinding alone in open circuit may be adequate. Otherwise, grinding followed by air classification is required. Figure 4.5.1 shows the general flowsheet illustrating the processing of high grade barytes.

A beneficiation plant with a yearly capacity 200,000 tonnes for processing low grade barytes is due to come up at a capital outlay of Rs.23 crores. A general flowsheet for processing of low grade barytes is shown in Figure 4.5.2. Carboxylic acids, alkyl sulphuric acids, and their salts are the collectors for flotation of barytes. Alternately, selective flotation of fluorite by depressing barytes can also be undertaken. Alkyl sulphate floats barytes. Desliming the ground feed prior to flotation helps in effective separation and reduction in quantities of reagents to be used. The iron oxide minerals associated with barytes can be selectively depressed at high alkalinity (11 pH) with a metasilicate (1 to 2 pounds /tonne ore).

In and around Mangampet village, several processing plants are in operation (5 to 6 TPH capacity x 18 hours/day = about 100 tonnes/day) producing finer than 95%, - 200 mesh size barytes for meeting ONGC’s oil drilling /export requirements. Besides the fineness (95% -200 mesh), ONGC specifies 4.2 sp. gr. of the ground product and below 0.05% water solubility for oil drilling use. Because of its high specific gravity, barytes, $BaSO_4$, is in great demand from the oil drilling industries.

Ore reserves of barytes as on 1st April 2010

Mineral	Reserves (million tonnes)	Production (million tonnes per annum)	Main producers
Baryte	72.734	2.334	A.P. Mineral Development Corporation Ltd. and Salaruddin Grey Barytes, Cuddapah (Seemandhra); Viswabharati Mining Corporation, Cuddapah (Seemandhra); Rajasthan Barytes Ltd., Udaipur (Rajasthan)

Source gratefully acknowledged: Statistical Profiles of Minerals 2010-2011 by Indian Bureau of Mines, Nagpur, April 2012.

Figure 4.5.1: Processing high grade Barytes

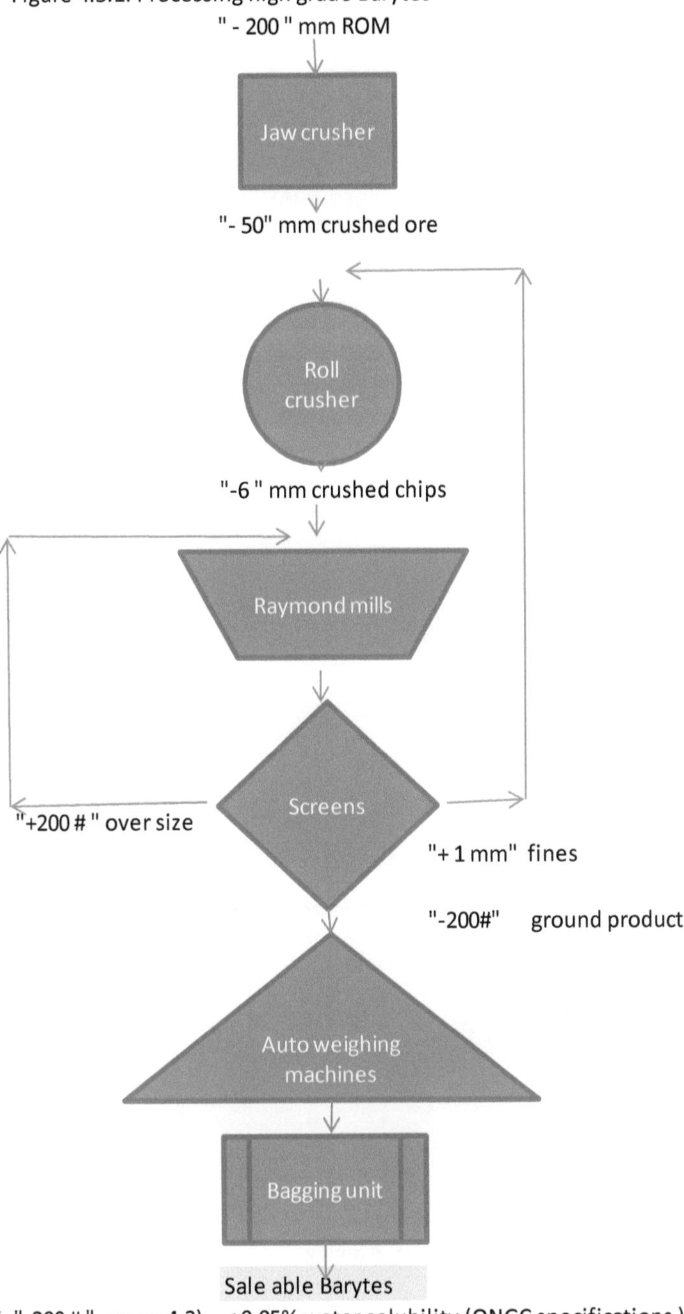

Ground barite (95% "-200 # ", sp.gr: 4.2) < 0.05% water solubility (ONGC specifications)

Figure 4.5.2: Beneficiation of low grade Barytes ore

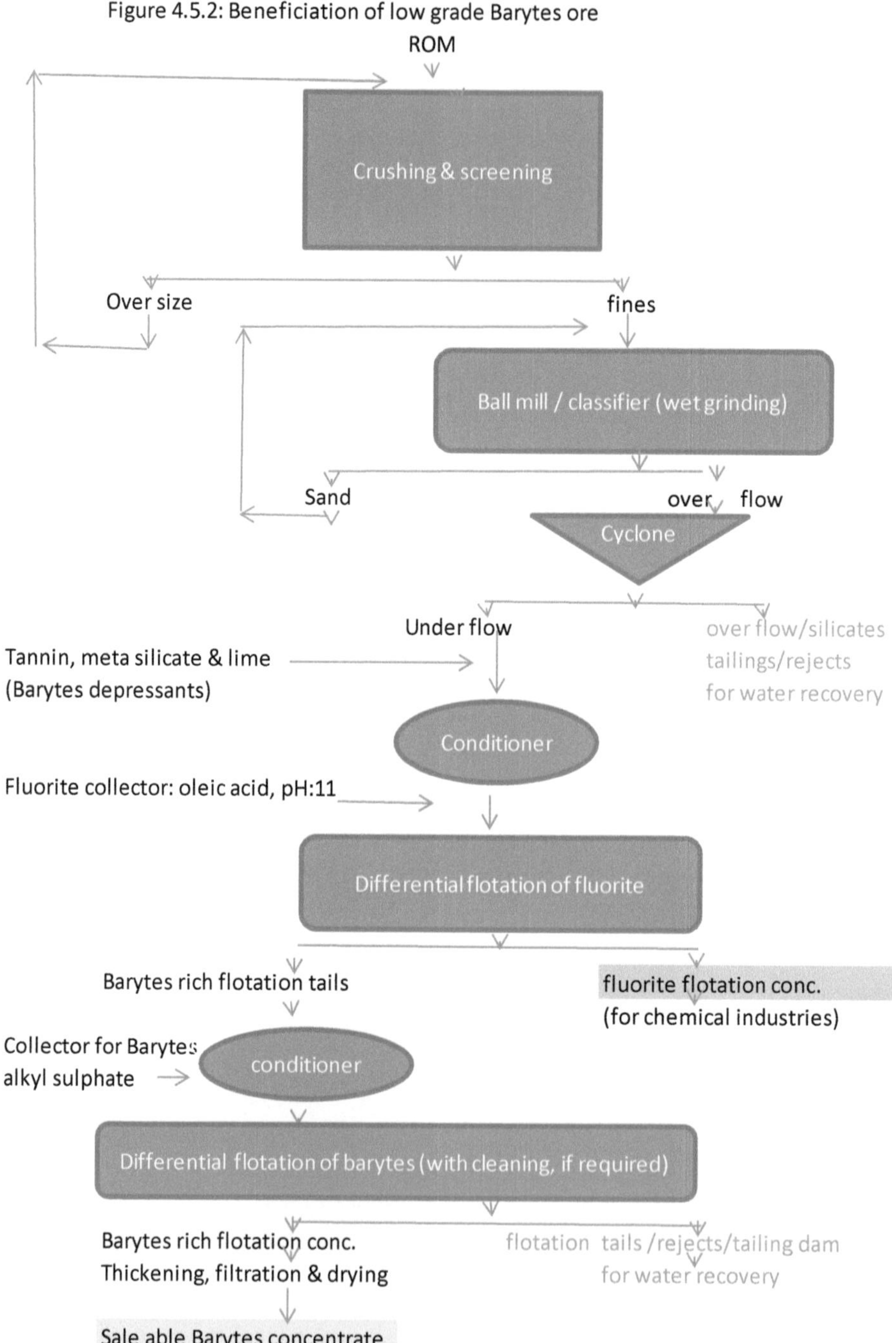

Chapter 5

Diamonds

Diamonds are found in kimberlite pipes. The incidence of diamonds available from Panna, Madhya Pradesh, is about 10 carats (equal to 2 grams by weight) per 100 tonnes of pipe rock (2 out of 10^8 concentration). For profitable commercial production, the incidence of diamonds should preferably be greater than 8 carats per 100 tonnes. Diamonds, located in kimberlite pipes of Panna, Madhya Pradesh, are commercially mined and recovered by NMDC Limited from its Majhgawan mine.

For keeping up its high market value for larger stones, diamonds have to be necessarily recovered with minimum possible breakage. About a decade back, subjecting the diamonds' bearing pipe rock for weathering at Panna for about 2 to 3 months helped the rock getting brittle which facilitated easy breakage (literally by hand) of rock and recovery of about 20,000 carats/per annum diamonds in their natural size. However, this operation is not universally practiced elsewhere in the world. Primarily for increasing the mine production many fold and due to space constriction this operation has been shelved now even at Panna.

Current practice. Figure 5.1 shows the current process adopted at Panna for recovery of diamonds. The pipe rock is crushed in a jaw crusher and then in a cone crusher working in closed circuit with a screen to liberation size of -20 mm. The -20 +1.25 mm crushed product is subjected to heavy media cyclones of 50 TPH capacity using ferrosilicon as the medium. The heavier sink product, which contains diamonds, is separated from the lighter float product (gangue). The sink product is dried, screened, and subjected to magnetic separation to recover ferrosilicon media for recycling. The closely sized sink fractions, free from lighter gangue and ferrosilicon, are subjected to X-rays sorting for recovery of most of the diamonds. The rejects/tailing portion are treated in grease tables to recover diamonds that might have slipped past the X-ray sorters. This flowsheet is adopted for larger production of diamonds, at about 1,00,000 carats per year. Dense media separation (DMS) followed by X-ray sorting and grease tables is more or less the method to be adopted for any future expansion in production of diamonds on a large scale.

With the introduction of DMS followed by mechanized X-ray sorters in the year 2003, diamond production has gone up to 65,000 carats/per year. However, due to delay in getting environmental clearances initially as well as subsequent delays in getting sanction of extension of expired mine lease, the current diamond production of NMDC out of its Panna mine is only 18,043 carats. They were sold at an average price of Rs.7,000/carat raw diamonds during the financial year 2011-2012.

Photo 5.1 shows the diamonds processing plant, including HMS operated by the NMDC Limited at Panna.

About a decade back, after mining and recovery of all the available diamonds in the adjacent Ramkheria mine of Panna, the mine was closed. Conglomerates of Racherla, in Kurnool district (Seemandhra), contain the diamonds. However, a detailed investigation carried out by NMDC Limited showed the incidence of diamonds available in the region to be low/or not economically viable and, hence, not enough for undertaking profitable commercial exploitation.

The value of diamonds depends on their cut, colour, clarity, and weight (carat). Their market value is assessed based on the following intrinsic characteristics of the diamond:

- Cut: The cuts are named as 'shallow' - ideal to get perfect double de-refraction (high-price diamond), and 'deep' - hearts and arrows (low price, value decreases). The price goes up with the perfect cut. Several firms in Surat developed the skill required for cutting diamonds perfectly.
- Colour: Colourless (high value diamond), near colourless, slightly tinted, very light yellow, light yellow, or brown (value decreases correspondingly).

- Clarity: High price for diamonds without inclusions; price decreases with inclusions (lower price).
- Carat: Value of diamond goes up with the weight of an individual piece in carats (1 carat = 200 mg).

Exploration works are in progress to locate diamond bearing kimberlite pipes. Both public sector companies (with assistance and expertise from the multinational companies) and private companies are engaged in locating the diamond-bearing kimberlite pipes that occur below the earth's surface. Diamond exploration is now confined to the following areas: Kalyandurg area in Anantapur district (Seemandhra) and Baghain, Sarang, Rampura, and Matwa in Madhya Pradesh.

Rio-Tinto has successfully located a large diamond deposit in Bunder area of Chhatarpur district. The firm is currently engaged in making a feasibility study for establishing a large-sized mine with a capacity to produce diamonds of more than 200,000 carats per annum.

Reserves and incidence of diamonds as on 1st April 2010

Mineral	Reserves (million carats)	Production (million carats)	Grade	Main producers
Diamond	31.922	0.019774	10 carats/ 100 tonnes of kimberlite pipe rock	Majhgawan, Panna (Madhya Pradesh) (NDMC Limited); 'Shallow (Uthali)' - Itwan, Panna, (Madhya Pradesh) Director of Geology and Mining, (Madhya Pradesh)

Source gratefully acknowledged: Statistical Profiles of Minerals 2010-2011 by Indian Bureau of Mines, Nagpur, April 2012.

Figure 5.1: Processing Kimberlite pipe rock for recovery of diamonds

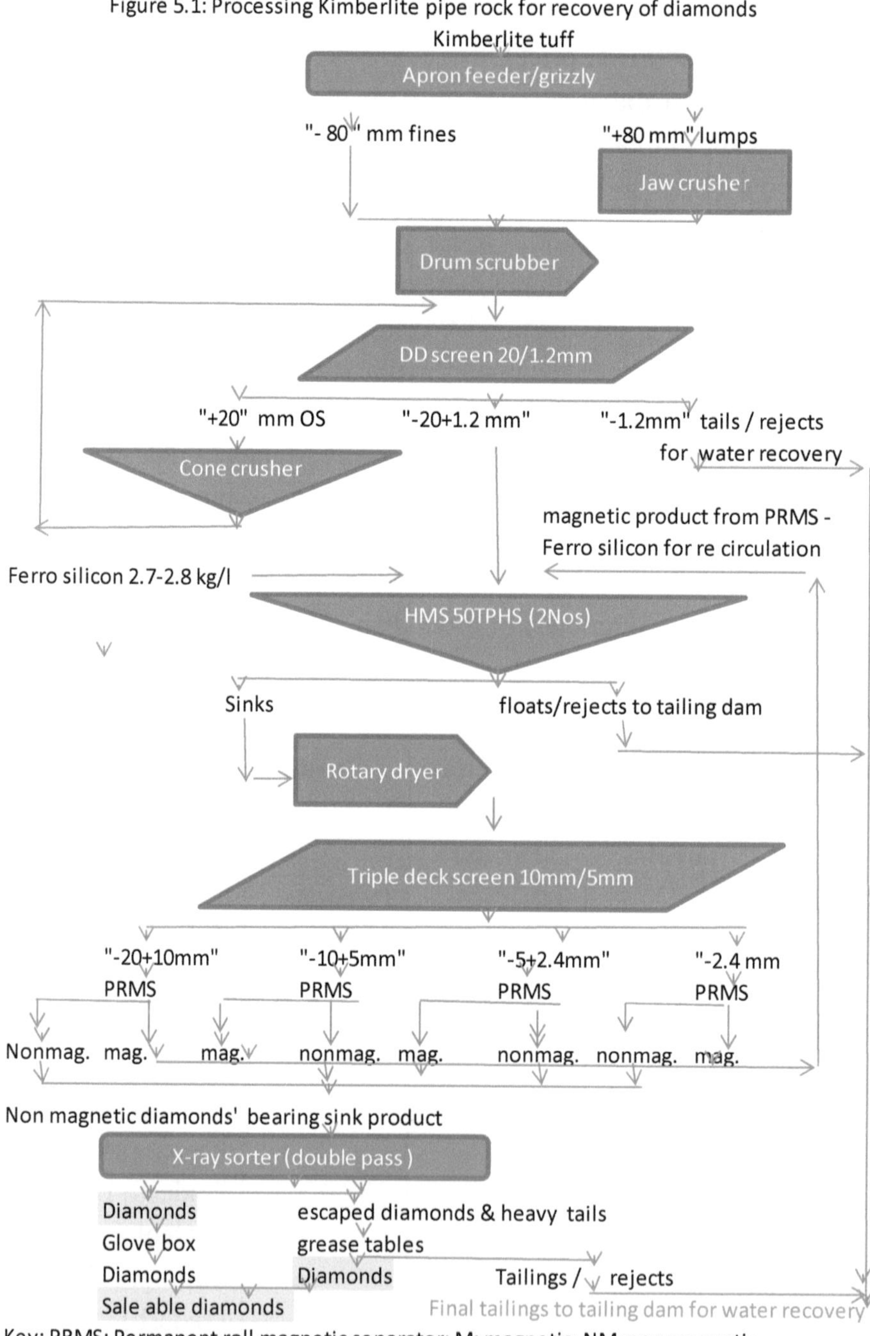

Key: PRMS: Permanent roll magnetic separator; M: magnetic; NM: non magnetic

Photo 5.1: View of Diamond-Mining Project, Panna, Madhya Pradesh

Courtesy: NMDC

Chapter 6

Base Metallic Ores (Copper, Lead and Zinc)

6.1 Copper ore

Chief ore minerals are chalcopyrite $CuFeS_2$, chalcocite Cu_2S, covellite CuS, malachite $Cu_2CO_3\,(OH)_2$, azurite $Cu_3\,(CO_3)_2(OH)_2$, native copper etc.

Beneficiation of low grade copper ore: Flotation is adopted for the beneficiation of low grade copper ore as per the general flowsheet shown in Figure 6.1

Flotation of fresh (un-oxidized) low grade copper, zinc, pyrite ores is undertaken with very small quantities of xanthates as collector, pine oil as frother, and cheaply available lime for pH control. Lime is effective to depress the associated pyrite with copper ores at more than 10 pH while selectively floating off copper minerals.

Pyrite associated with copper ores may contain very minor quantities of precious metals like gold and silver. The gold is finely interlocked with pyrite. Recovery of gold involves bulk flotation of copper minerals along with pyrite leading to poorer grade of copper concentrate. The precious metals are recovered during copper smelting.

Beneficiation of copper ore (chalcopyrite containing about 0.8–2% Cu) is carried out at Ghatsila, in Bihar, and Khetri in Rajasthan. The chalcopyrite ore at Ghatsila contains precious metals interlocked with pyrite. Sphalerite flotation is commercially undertaken at Zawar, Rajasthan.

Beneficiation is aimed to produce copper concentrate of about 20-25% Cu, with recovery of more than 90% copper metal.

The flotation reagents per tonne of ROM include use of about 0.05 kg/tonne potassium ethyl/amyl xanthates, about 0.05 kg/tonne pine oil, and 1-2 kg/tonne lime.

In copper circuit, about 0.5 kg/tonne sodium sulphide is used to depress sphalerite.

In zinc circuit, about 0.35-0.50 kg/tonne copper sulphate is added as an activator to float sphalerite.

Reserves of copper ore as on 1st April 2010

Mineral	Reserves (million tonnes)	Production (million tonnes)	Grade % Cu	Main producers
Copper ore Copper metal	1558.458 12.287	3.615 0.137 concentrate	0.88 to 1.0	Malanjkhand (Rajasthan); Khetri and Kolihan in Jhunjhunu (Rajasthan); Surda in Singhbhum, (Jharkhand) (all of Hindustan Copper Ltd.)

Source gratefully acknowledged: Statistical Profiles of Minerals 2010-2011 by Indian Bureau of Mines, Nagpur, April 2012.

Figure 6.1: Beneficiation of copper and zinc ores

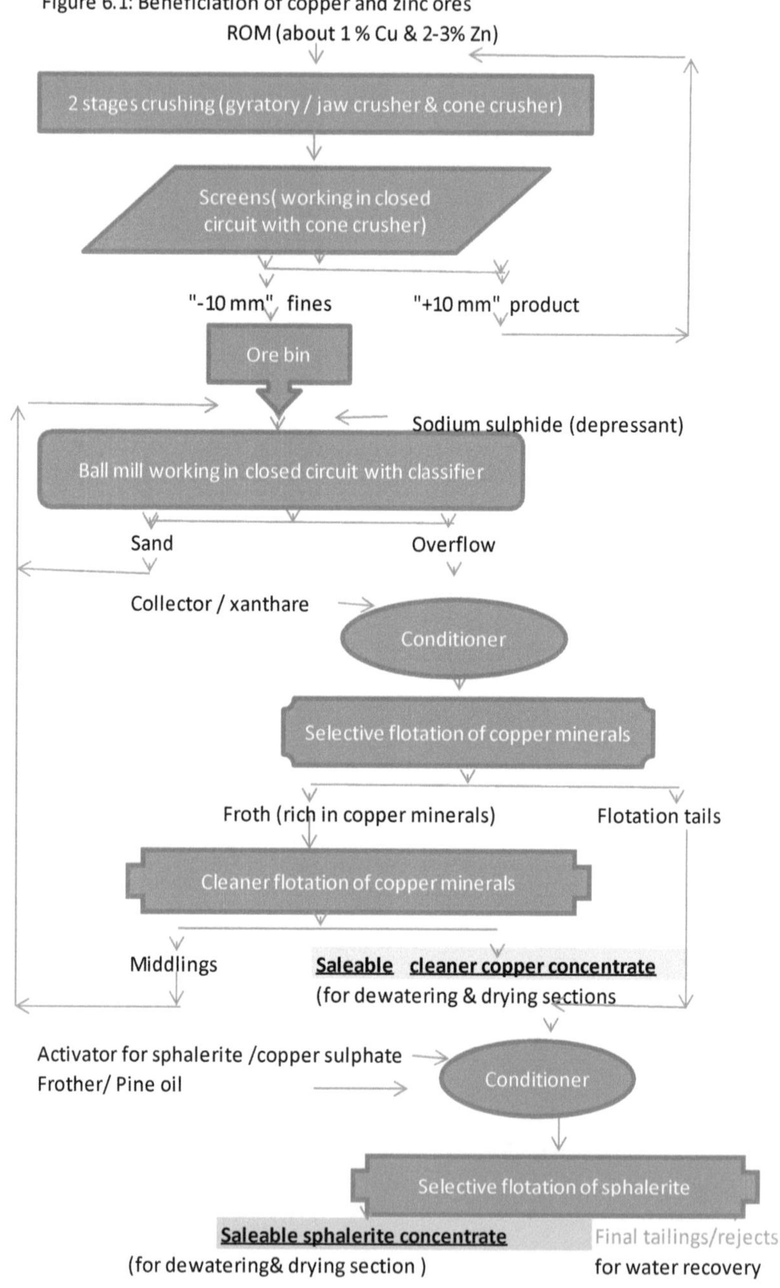

6.2 Lead and zinc ore (galena and sphalerite)

Lead minerals: The lead ores are galena (PbS), cerussite ($PbCO_3$), and anglesite ($PbSO_4$).

Galena, an important mineral, is often associated with sphalerite, silver, and pyrites. Galena can successfully be separated from sphalerite by differential flotation, as shown in Figure 6.2, producing a concentrate of 65 to 80% Pb. Due to intimate association, galena floats together with silver. Lead is smelted together with silver but finally separated only during the refining stage.

Zinc minerals: Important zinc ores are sphalerite (ZnS), marmatite ($(ZnFe)S$. zincite (ZnO), Smithsonite ($ZnO.CO_2$), and franklinite ($ZnFe_2O_4$). Sphalerite and marmatite, the main economic ores, are usually associated with galena. They can easily be separated by differential flotation producing a concentrate containing 50 to 60% Zn.

Reserves, grade, and chief producers of lead and zinc ores as on 1st April 2010

Mineral	Reserves (million tonnes)	Production (million tonnes)	Grade %	Main producers
1. Lead and zinc ore	685.595	7.49	Pb:2.03 and Zn:11.54	Rampura Agucha; Sundesar Khurd and Balaria, all in Udaipur (Rajasthan); Rajpur Dariba, Rajsamand, (Rajasthan) under the control of Hindustan Zinc Ltd., Rajasthan
2. Lead metal	11.549	Lead concentrate: 0.145	Pb: 57.46% (concentrate)	
3.Zinc metal	36.665	Zinc concentrate: 1.42	Zn: 51.16% (concentrate)	

Source gratefully acknowledged: Statistical Profiles of Minerals 2010-2011 by Indian Bureau of Mines, Nagpur, April 2012.

Figure 6.2: Beneficiation of lead & zinc ore

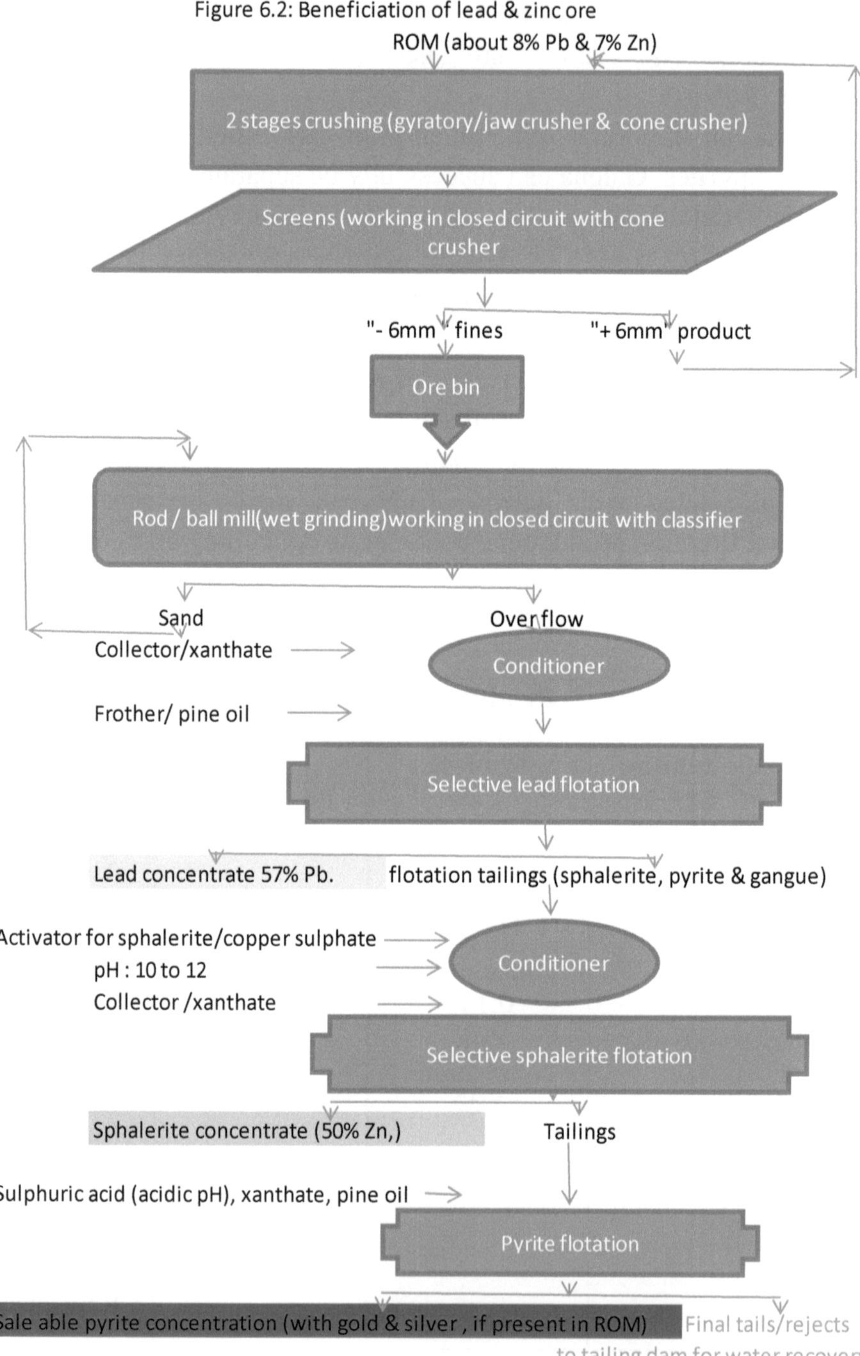

6.3 Oxidized base metal ores of copper, lead and zinc

The prominent carbonates of copper ore are: azurite $2CUCO_3 .Cu(OH)_2$ and Malachite $CuCO_3.Cu(OH)_2$ which are associated with quartz, clays, silicates, and hydrated iron oxides.

Carbonate of lead ore: Cerussite $PbCO_3$, is associated with gangue containing hydrated iron oxides, clays, silicates, and other carbonates.

Carbonate of zinc ore: Smithsonite $ZnO.CO_2$ is associated with hydrated iron oxides, silica, and clays.

Beneficiation of these oxidized minerals depends on the type of associated gangue minerals, namely, siliceous, carbonate, clayey, or ferruginous types.

Three general methods are used for beneficiation of these oxidized ores:

1. Sulphodize the desired carbonate mineral with hydrogen sulphide or sodium sulphide, and float it like sulphide minerals with amyl xanthates.
2. Use carboxylic acids as collectors to float base metal carbonates while depressing gangue like silicates and clay. However, this method is not effective to depress ferruginous and carbonate gangue. Although it is easy to float the desired mineral, it is not easy to prevent flotation of unwanted gangue, which hinders the separation process.

Float base metal oxides using sulfhydryl collectors containing more than 4 to 6 carbon atoms such as xanthates and dithiophosphates but with less selectivity of the associate gangue.

Chapter 7

Ferro-alloying Ores (Manganese, Chromium, Nickel and Molybdenum)

7.1 Manganese ores

Pyrolusite (MnO_2), Manganite (Mn_2O_3), Psilomelane ($MnO_2.H_2O.K_2BaO_2$), and Braunite ($3Mn_2O_3.MnSiO_3$) are some of the common manganese ores. These are available as surface outcrops. Chief impurities are alumina, silica, and iron oxides.

Amenability of the ore for beneficiation: As market price of these ore depend on their manganese content, beneficiation of low grade ores is carried out to improve metal content in the ore. The separation technique includes removal of clay in scrubbers/log washers, crushing the ore free from clay, followed by gravity concentration using jigs and tables. Gravity concentration can improve the ores quality by just 1 to 2% due to inter locking/intimate association of the associated gangue with the manganese minerals. Ores intimately associated with silicates and oxides are not easily amenable for beneficiation. However, medium grade manganese ores are mostly treated by the cheaper gravity methods of concentration as shown in Figure 7.1.1. In case the manganese ore is a silicate, carbonate, or oxide ore, ultrafine grinding is required for their beneficiation. In a few cases, they are processed by flotation using fatty acids. Flotation is resorted to improve the quality of ore containing <20% Mn, using soaps and oil reagents. The flotation concentrate is calcined and nodulized to produce a saleable product having >48% Mn. Intimate association of iron and silica with manganese mineral in ferruginous manganese ores makes it difficult to separate gangue minerals by flotation using fatty acid reagents. Therefore, the ore has to be beneficiated by gravity concentration followed by reduction roast technique as shown in Figure 7.1.2.

Depending on its grade and quality/purity, manganese ore is used by the appropriate industry meeting its quality requirements.

High grade ores are used for making batteries.

Ores containing >45 % Mn with Mn to Fe ratio of more than 7:1 are used for making ferro-alloys.

Ores with Mn to Fe ratio of less than 7:1 are used for spiegeleisen.

Ores containing < 12% Mn are used for making manganese pig iron.

Reserves, production, grade, and main producers of manganese ore as on 1st April 2010

Mineral	Reserves (million tonnes)	Production (million tonnes)	Grade %	Main producers
Manganese ore	429.980	2.881	>50% ; current production: 25 – 35% Mn	Balaghat (Madhya Pradesh) and Dongribuzurg (Maharashtra) (under the control of MOIL); Shivrajpur (Gujarat) (under the control of GMDC); Patmunda in Sundergarh (Orissa) (under the control of OMM, Orissa)

Source gratefully acknowledged: Statistical Profiles of Minerals 2010-2011 by Indian Bureau of Mines, Nagpur, April 2012.

Figure 7.1.1: Processing pyrolusite ores

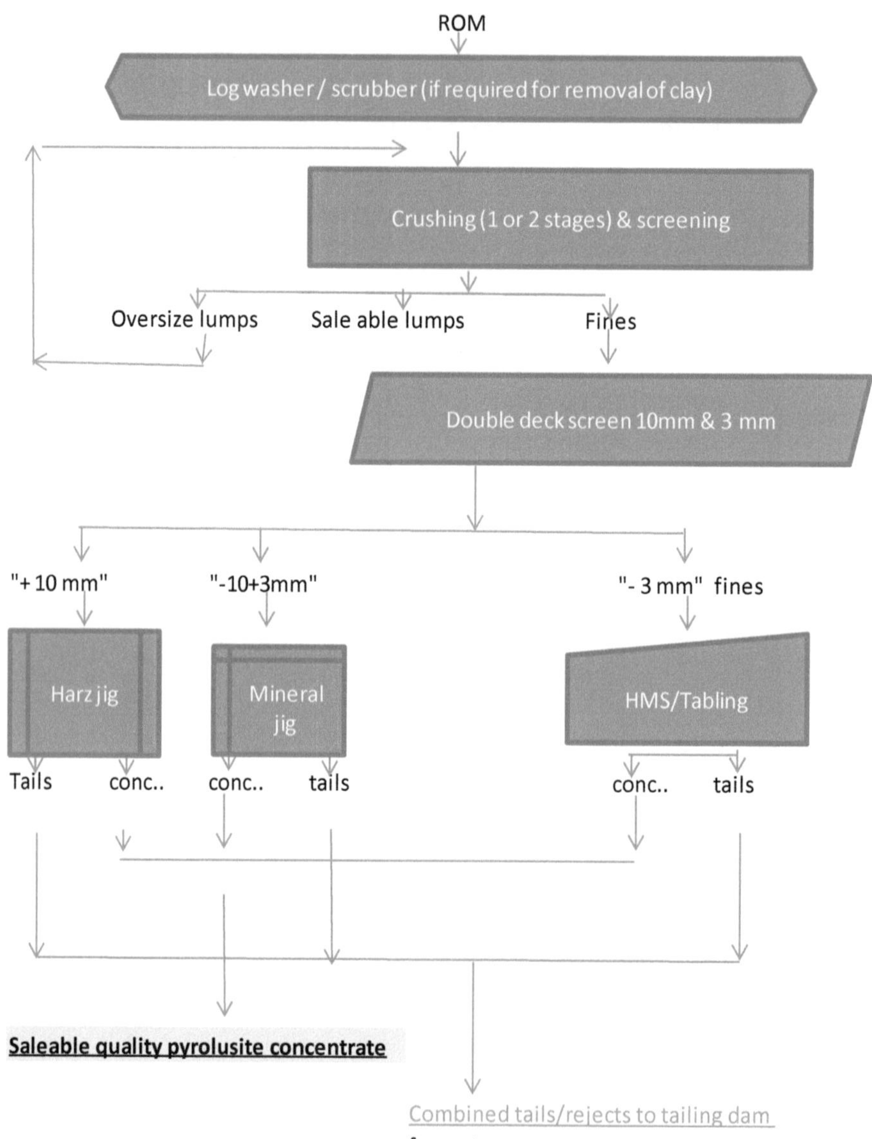

Figure 7.1.2: Beneficiation of ferruginous manganese ores

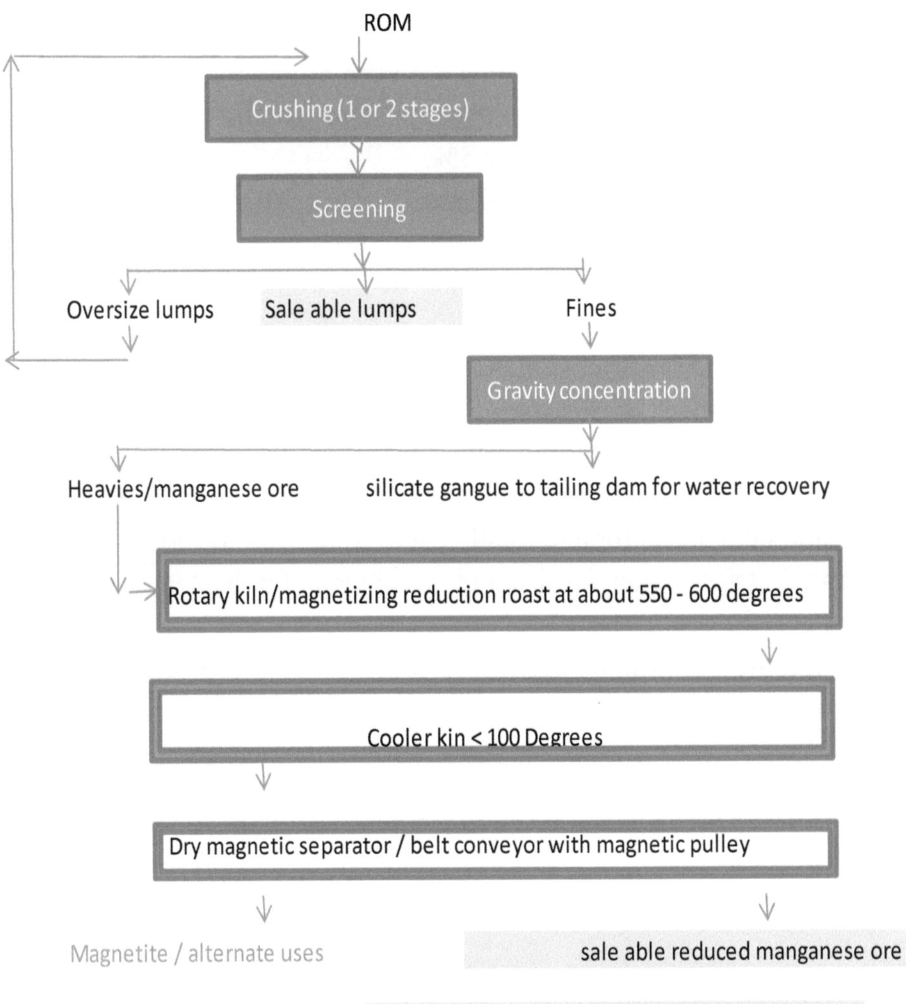

7.2 Chromite ore

Chromite (FeO Cr_2O_3) is the only source of chromium.

Beneficiation of chromite: An important criteria to be kept in view during beneficiation is to maintain the ratio of chromium to iron to 3:1 for use by the ferro-chrome industries. The market specifications are Cr_2O_3: 48%, S <0.05% and silica 5%.

Many chromite ores are marketed directly after mining, crushing, and hand sorting.

Gravity methods of concentration of medium grade chromite ores, including jigging or tabling, are common as shown in Figure 7.2.1.

Beneficiation of low grade chromite ores as shown in Figure 7.2.2 by gravity methods of concentration is not really helpful as it does not improve the chromium to iron ratio. Ores associated with magnetite are therefore troublesome. Removal of magnetite by magnetic separation will not help as chromite totally free from iron is not acceptable. Likewise, excess iron is also not acceptable for heat and corrosion resistant refractory uses.

Chromite is used in metallurgical (ferro-chrome), chemical, and refractory industries.

Reserves, production, grade, and main producers of chromite ore as on 1st April 2010

Mineral	Reserves (million tonnes)	Production (million tonnes)	Grade % Cr_2O_3	Main producers
Chromite	203.346	4.262	Mostly <52% (both lumps and fines); less than 18% quantity of fines produced are >52%	Sukinda (Odisha) (by TISCO) and Kaliapani of Jaipur by OMC, Indian Metals & Ferro Alloys Ltd., and Balasor Alloys Ltd.

Source gratefully acknowledged: Statistical Profiles of Minerals 2010-2011 by Indian Bureau of Mines, Nagpur, April 2012.

Figure 7.2.1 : Processing chromite ores

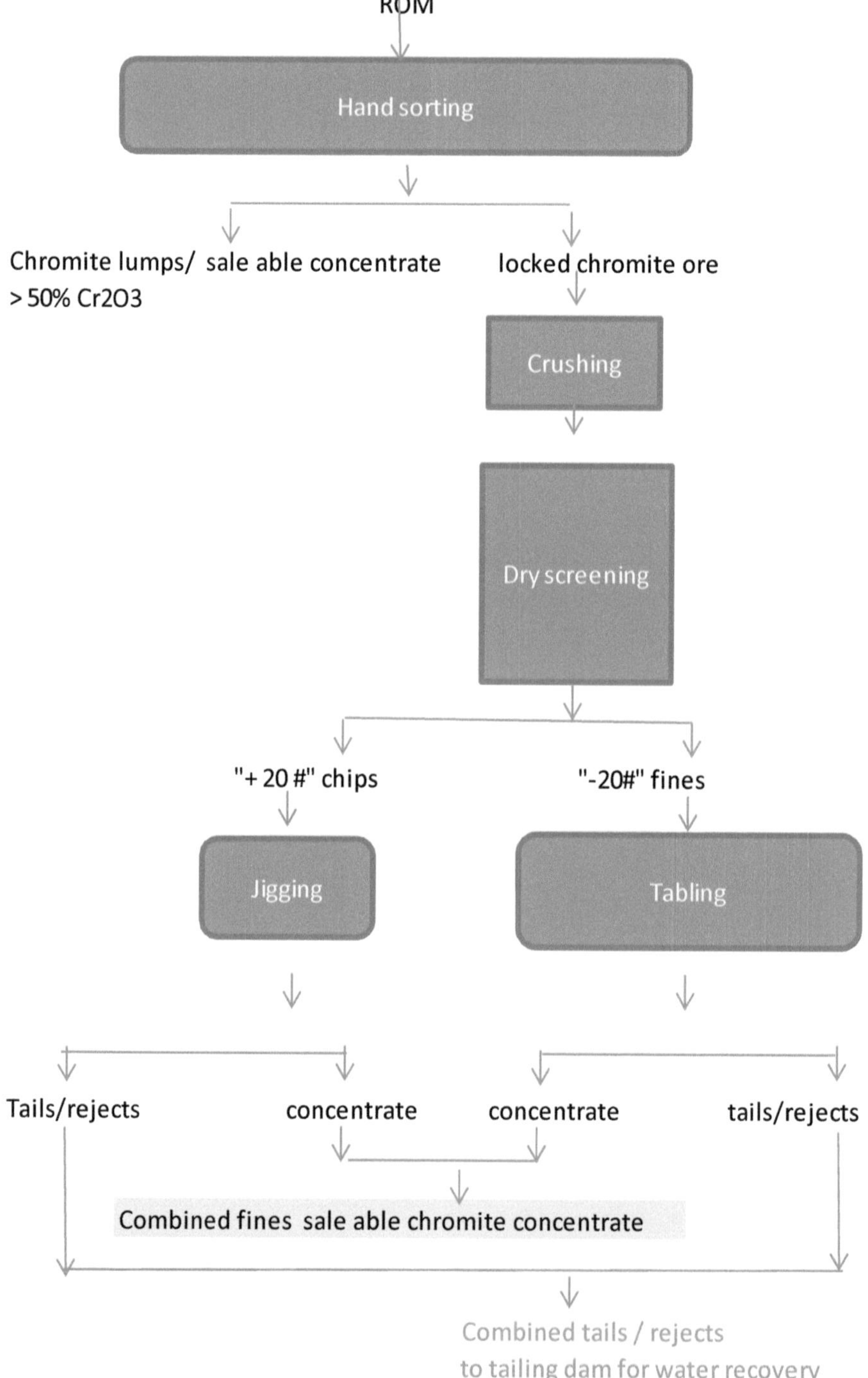

Figure 7.2.2: Beneficiation of low grade chromite ores

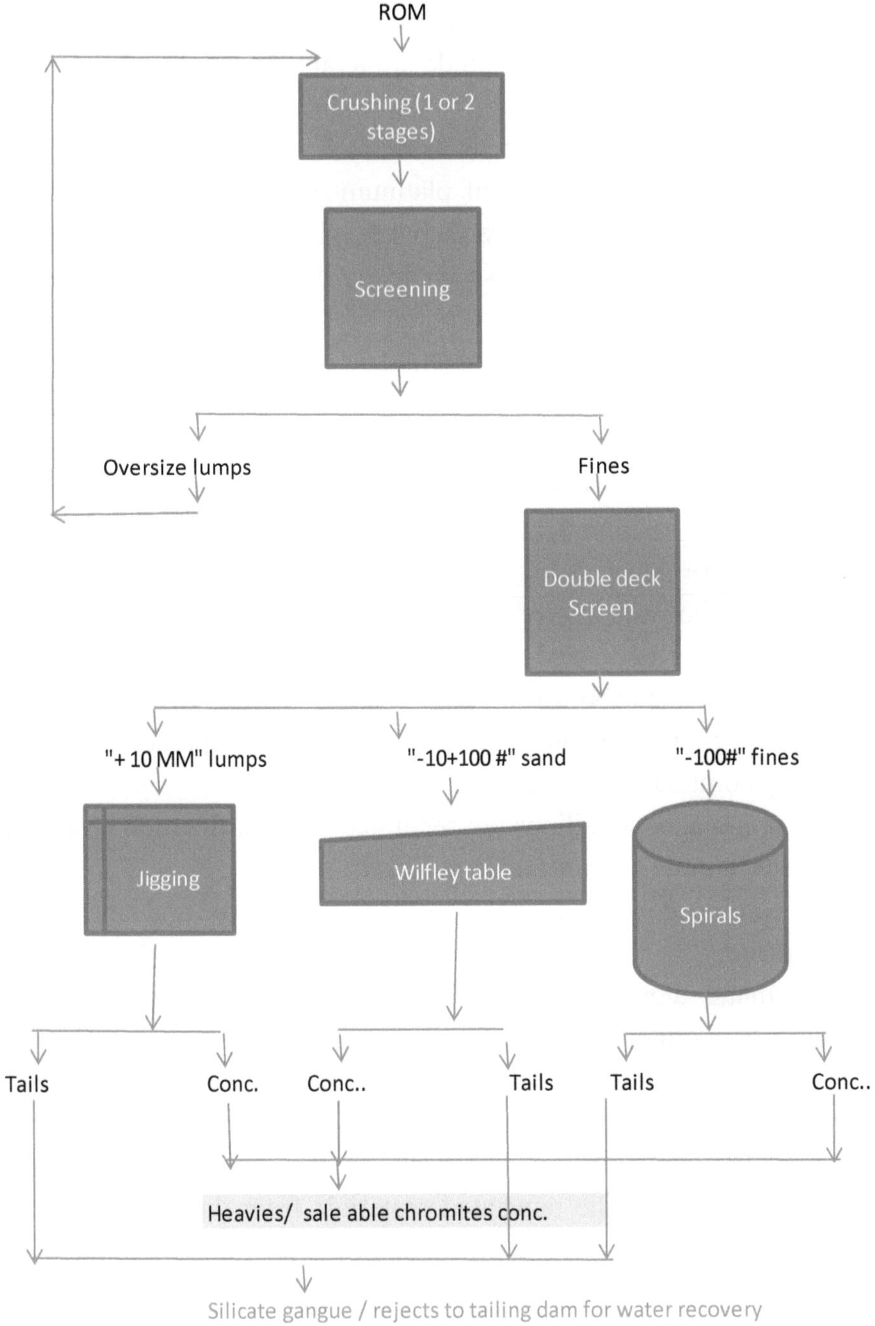

ROM

Crushing (1 or 2 stages)

Screening

Oversize lumps

Fines

Double deck Screen

"+ 10 MM" lumps

"-10+100 #" sand

"-100#" fines

Jigging

Wilfley table

Spirals

Tails

Conc.

Conc..

Tails

Tails

Conc..

Heavies/ sale able chromites conc.

Silicate gangue / rejects to tailing dam for water recovery

7.3.1 Beneficiation and extraction of nickel ore

The primary nickel ores are sulphides and arsenide. Millerite NiS and Pentlandite (Fe Ni) S are sulphide minerals. Niccolite NiAs is less common.

Pentlandite, the nickel ore, is associated with chalcopyrite ($Cu Fe S_2$), cobalt, silver, and metals of platinum group. The world's largest Canadian deposit contains more than 1% Ni. However, in India, some of the copper sulphide ores (like chalcopyrite ore of Ghatsila) contain small amounts of nickel sulphide (about 0.1% Ni), besides molybdenum, gold, silver, and cobalt.

Principle of nickel extraction: In the absence of mining and extraction of major nickel deposit in India, the practice adopted by the International Nickel Co. of Canada, which produces almost the whole world's nickel requirement from its large underground mine of Pentlandite ore located in Sudbury basin[1] essentially by flotation technique, is shown in Figure 7.3.1.

Pentlandite is associated with chalcopyrite, besides other metals.

The ore contains about 1 to 1.7 % nickel and around 1% copper.

Pentlandite is separated by using magnetic separation process, producing combined nickel-copper concentrate.

The non magnetic tailings are separated by bulk flotation producing copper concentrate and rejectable tailings.

The combined nickel copper concentrate is sintered and smelted in a blast furnace.

The matte along with nickel copper concentrate is charged to Bessemer converters.

The Bessemer matte is calcined, and copper is removed by leaching process. The nickel is recovered by electrolysis.

Thus, the metallurgical treatment of the final dry concentrate involves slagging of the unwanted materials from the enriched matte containing nickel, copper, and precious metals for further treatment and recovery.

---

[1] Dennis, W.H. (1953). *Metallurgy of the nonferrous metals*. London: Pitman.

Flotation reagents used in flotation circuit: After crushing and grinding the ore to its liberation size, nickel sulphide is floated along with copper sulphide using xanthates, pine oil, and lime (bulk flotation). In the next step, differential flotation is adopted, using sodium sulphide as a depressant to separate copper sulphide from nickel sulphide. Bulk flotation followed by selective flotation produces a copper concentrate, a nickel-copper concentrate, and a rejectable tailing.

Uses: Nickel is the most important component of many alloys, both ferrous and nonferrous. It is also used in making coins, batteries, electroplating etc.

Reserves of nickel ore as on 1st April 2010:

Nickel is found in Cuttack and Mayurbhanj, both in Odisha. There are no sizable reserves. Our country has to depend on imports.

Figure 7.3.1: Separation of nickel sulphide from copper sulphide & extraction of nickel

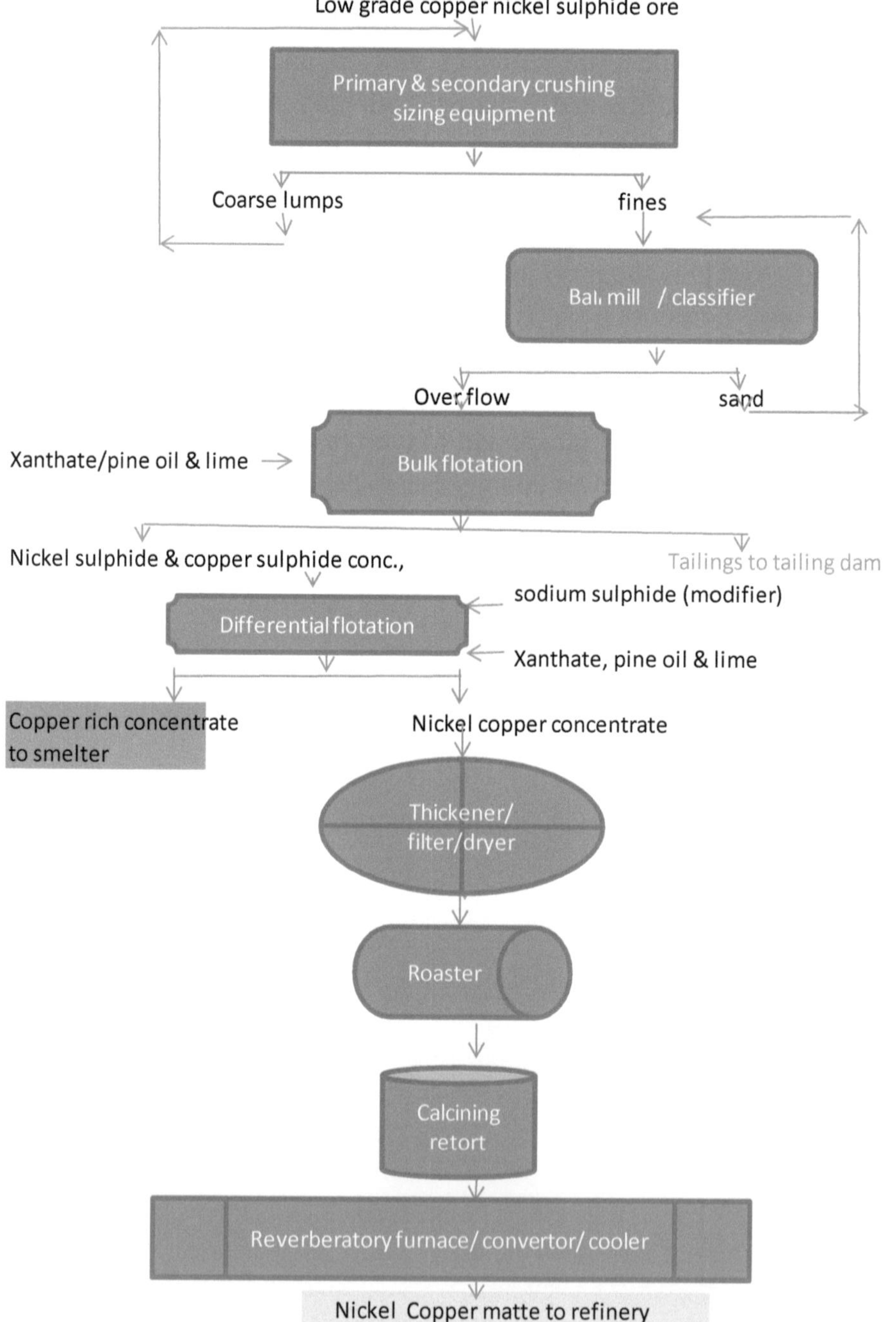

7.3.2 Nickeliferus pyrrhotite

Nickel occurs with nickeliferus pyrrhotite, chalcopyrite, gold and silver. Use of an alternate processing route shown in Figure 7.3.2. depends on the associated minerals and liberation size.

Figure 7.3.2: Beneficiation of low grade nickeliferus pyrrhotite

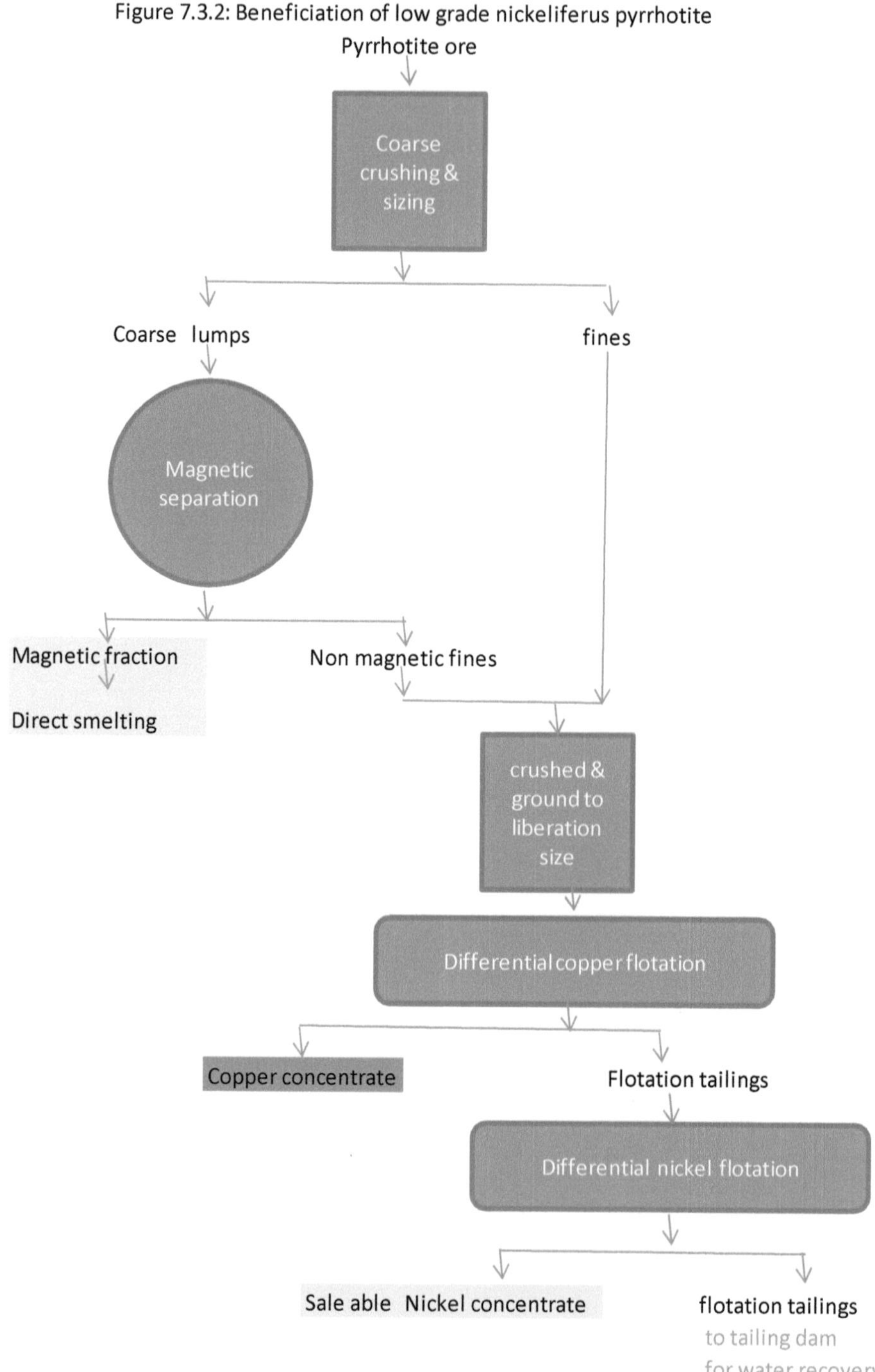

7.4 Molybdenite

Molybdenite (MOS_2) is the chief ore of molybdenum and found mostly in acid igneous rocks like granites and pegmatite.

Occurrence: Molybdenite is available in very small concentrations (<0.1% Mo). This is finely disseminated in copper sulphide ores of Ghatsila. It is recovered as a byproduct in the concentration of chalcopyrite ore. It occurs at Velampatti, Kattur, and Sundangipatti in the Harur belt of Dharmapuri, Tamil Nadu. Tamil Nadu Minerals are entrusted with the exploitation of the deposit.

Beneficiation technique: Molybdenite is a soft, shiny, blue sulphide mineral closely resembling graphite. Once molybdenite gets liberated from the associated minerals, it is easily recoverable by differential flotation using fuel oil/kerosene as collector and pine oil as frother (sometimes pine oil alone is good enough). Due to its flaky nature, gravity methods are not suitable for its concentration. On the other hand, it is easily floatable. However, to achieve the required marketable grade, a rougher concentrate produced at a coarser grind (say -100#) is reground to its liberation size and refloated/cleaned repeatedly (say twice or thrice) as shown in Figure 7.4 to achieve the marketable grade. The associated copper sulphide from the molybdenite concentrate can be removed by roasting it at 2500°C for conversion of copper sulphide to copper oxide prior to floating off molybdenite using kerosene-pine oil. In the alternative, the flotation reagents (both adsorbed and free layers) have to be totally removed from the thickened pulp of molybdenite-copper sulphide concentrate by steam heating to boiling temperature and holding it for a couple of hours. After cooling, molybdenite is floated off from the combined concentrate free from reagents using pine oil-kerosene combination. As molybdenite is easily floatable, it is difficult to depress it while selectively floating off copper sulphide minerals. However, it is possible to depress molybdenite using soluble starch and float copper sulphide. This results with a copper float free of Molybdenite and of Molybdenite-rich non float. The non float is subjected to differential flotation floating off molybdenite using pine oil-kerosene combination while depressing copper minerals.

Concentrate grade requirement: Alloy industries require >85% molybdenite. Copper is an undesirable impurity and needs to be separated by repeated cleanings. In the alternative, copper associated with the molybdenite concentrate has to be leached out chemically to meet the market specifications.

Molybdenum is required by the defence and aeronautical industries.

Reserves of molybdenum as on 1st April 2010

Mineral	Reserves (million tonnes)	Occurrences	Grade (% MO_2)	Main producers
Molybdenum ore	19.29	Harur belt in Dharmapuri District; A. Velampatti, Kattur and Sundangipatti in Harur Belt (Tamil Nadu)	0.102	Tamil Nadu Minerals are working on development of these occurrences
Molybdenum sulphide	0.012640			

Source gratefully acknowledged: *Indian minerals yearbook* (2011, part II, 50th ed.) , October 2012.

Figure 7.4: Beneficiation of molybdenite

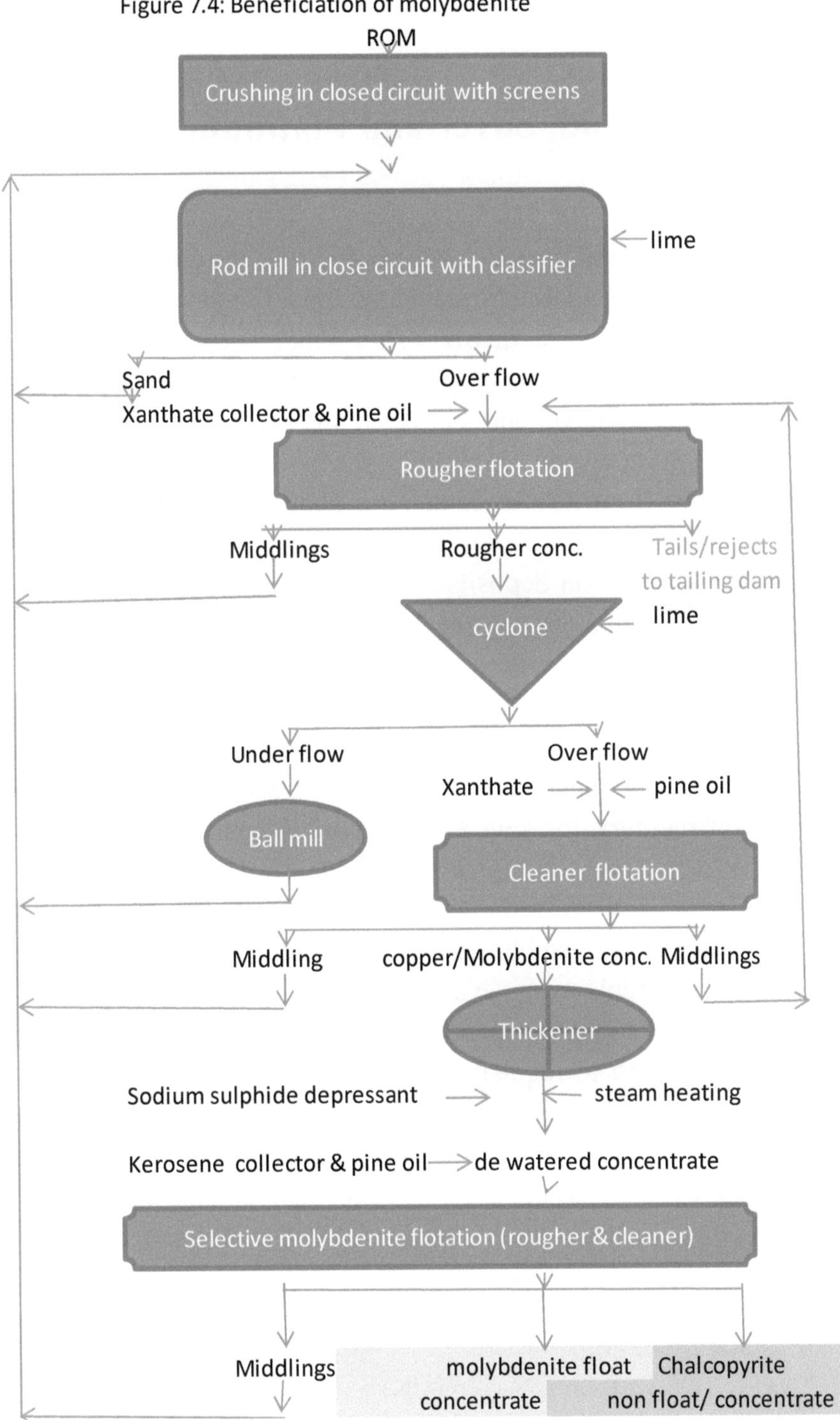

Chapter 8

Extraction of Precious Metals (Gold, Silver and Platinum)

8.1 Extraction of gold ores by cyanidation and precipitation of gold

Principle: Gold and platinum are commonly found in metallic form. Gold generally occurs in quartz veins, placer deposits, and usually associated with sulphides (pyrite, chalcopyrite, galena, arsenopyrites). Gold is alloyed with silver and recovered together.

Tenor: With the current high gold prices, it is commercially viable to recover gold from an ore containing about 1 to 2 grams gold per tonne ore.

Native gold in placer deposits/river sands/sediments is recovered by panning, sluicing etc., utilizing the high specific gravity of gold (15.6 to 19.3). Gold associated with sulphides is enriched by flotation using xanthates.

Earlier, enriched gold in the ore was used to be recovered by amalgamation. This technique involves allowing gold coming into contact with mercury forming an amalgam. At periodic intervals, the amalgam is squeezed through a cloth/hydraulic press to remove the excess mercury. The pressed amalgam containing gold and mercury is distilled. The mercury sublimes leaving behind the gold. Now, this practice has been replaced by cyanidation method to recover gold.

Cyanidation is the widely used method adopted for recovery of gold from the ore ground to requisite fineness as shown in Figure 8.1.1. In presence of oxygen, gold, dissolved in dilute solution of sodium cyanide in alkaline medium (lime is added), is recovered through its precipitation with zinc as per the following chemical reactions:

Dissolution of gold:

$$2\,Au + 4\,Na\,CN + O + H_2O = 2\,Na\,Au\,(CN)_2 + 2\,Na\,OH$$

Precipitation of gold:

$$2\,Na\,Au\,(CN)_2 + Zn = Na_2Zn\,(CN)_4 + 2\,Au$$

The main steps in cyanidation are as follows:

1. Size reduction of gold ore is carried out by crushing, grinding, and cycloning to its liberation size. Grinding is usually carried out in the presence of cyanide solution to ensure dissolution of most of the gold.
2. In the presence of oxygen and in alkaline medium, gold is dissolved in dilute cyanide solution.
3. Separation of the gold-bearing pregnant solution from the waste solids using thickeners.
4. Precipitation of gold out of the gold-bearing pregnant solution by zinc dust.

Figure 8.1.2. shows the alternate flowsheet for gold recovery from the cyanide solutions by carbon in pulp technique.

Reserves, production, grade, and main producers of gold as on 1st April 2010

Mineral	Reserves (million tonnes)	Produc-tion (tonnes)	Grade	Main producers
Gold primary ore	519.816	727000	3.07 gr	Hutti, Uti, and Hirabuddin, all in Raichur, Karnataka, under the control of Hutti Gold Mines Ltd.; Kunderkocha of East Singhbhum, under the control of Manmohan Industries (P) Ltd.
Gold metal	0.000666	2.239	grams/tonne	

Source gratefully acknowledged: Statistical Profiles of Minerals 2010-2011 by Indian Bureau of Mines, Nagpur, April 2012.

Figure 8.1.1: Beneficiation of gold bearing low grade ore & extraction of gold

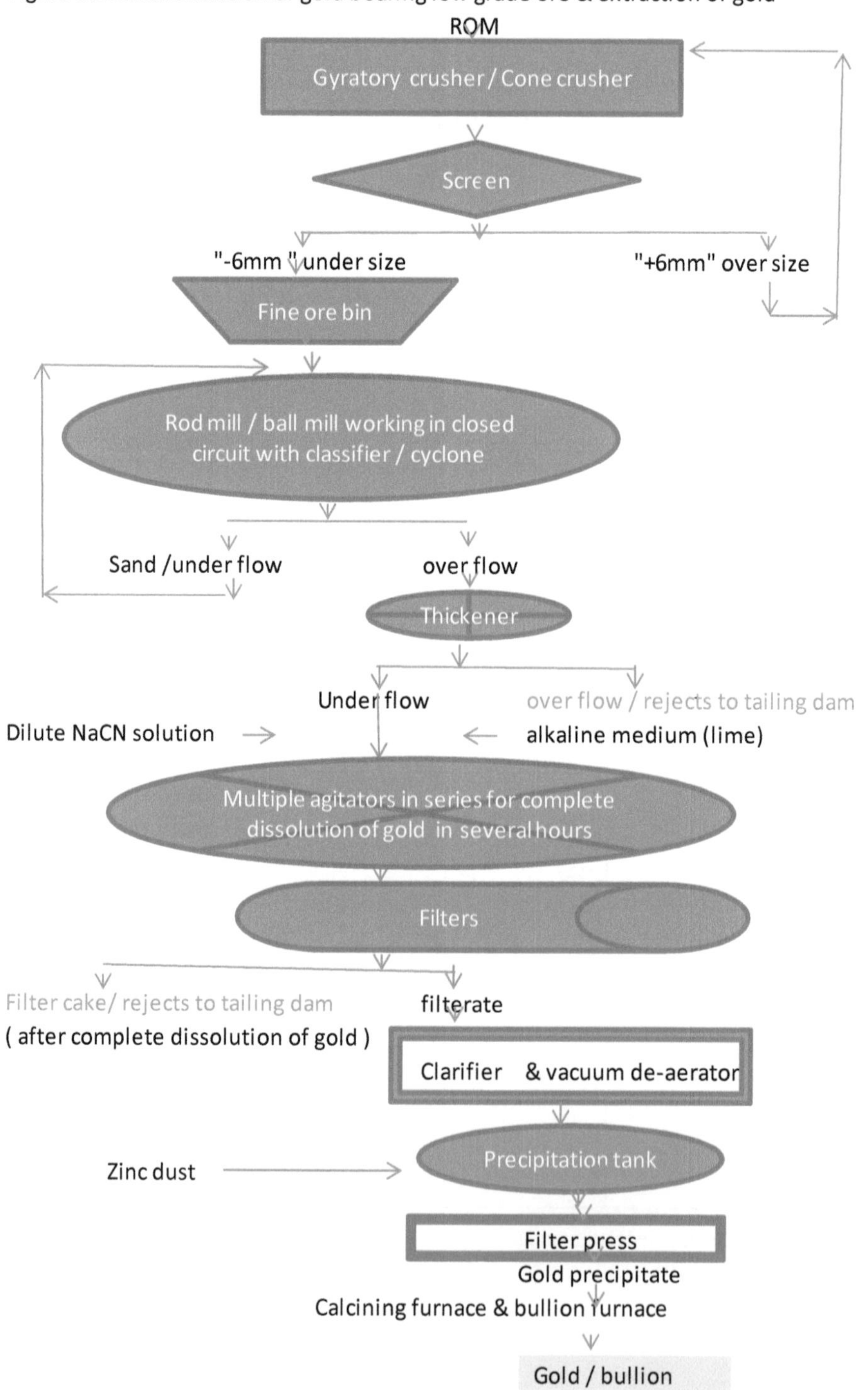

Figure 8.1.2: Beneficiation of gold bearing low grade ore & extraction of gold (C I C process)

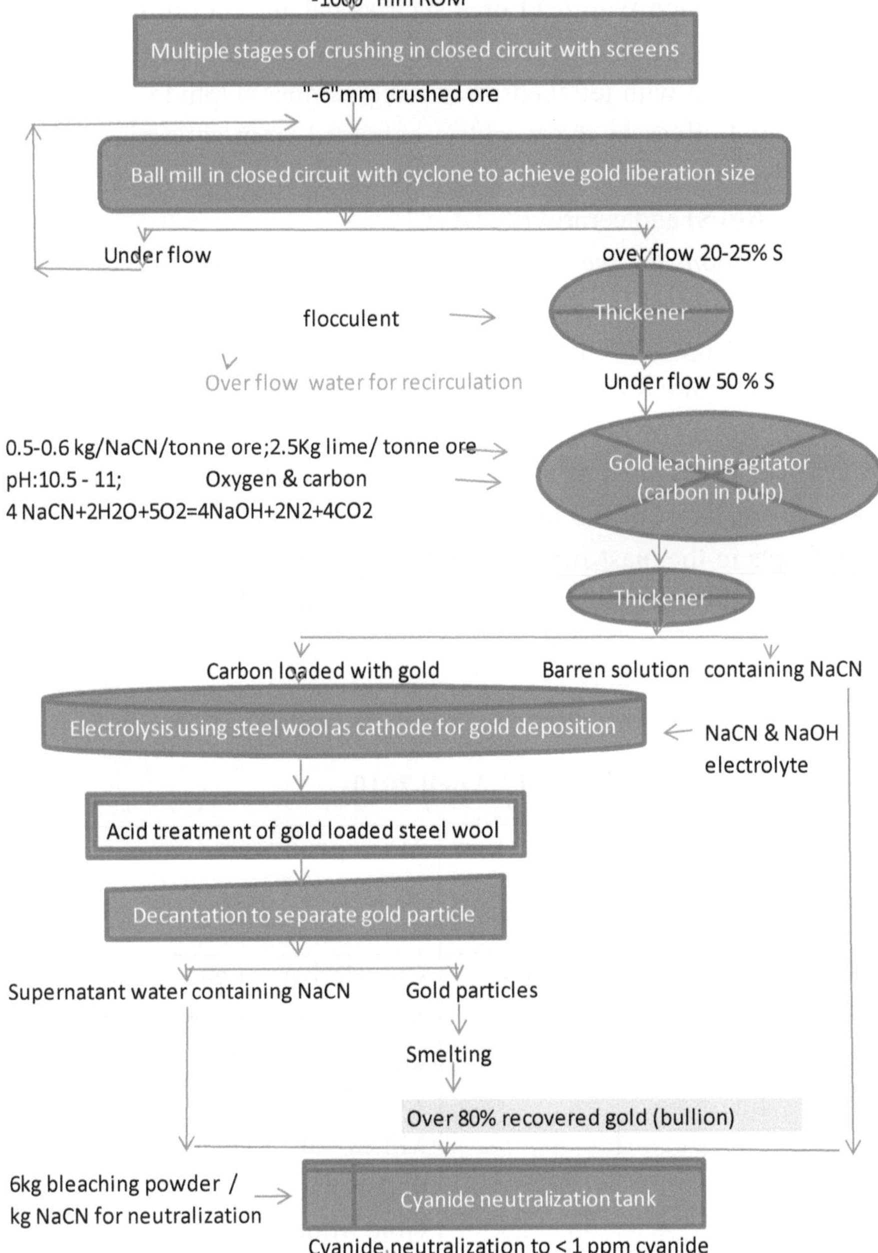

MSPL's gold project at Gadag flowsheet: gratefully acknowledged.

8.2 Silver

Silver is associated with lead ores such as galena, tin ore (cassiterite), and cobalt ores. It also occurs as an alloy with metallic gold, native copper ores, and with tetrahedrite in many copper sulphide ores. About 75% of total silver in the world is extracted from refineries of base metals. The rest 25% silver is produced from silver ores such as argentite (Ag_2S) and cerargyrite ($AgCl$).

Beneficiation of silver ore: Like gold, silver is also extracted from cyanide leaching of the ore ground to the requisite fineness but only after 80 hours agitation time followed by its precipitation with zinc dust as shown in Figure 8.2.

Silver as a by-product from refineries of base metals (lead ore): Silver is obtained as a by-product in the treatment of base metals like lead ore. Most of the lead ores contain silver. During the smelting of lead ores in the blast furnace, zinc metal is added to extract silver as zinc-silver alloy. As silver is lighter than lead, it floats on the molten lead. The zinc-silver is recovered by skimming. By heating, silver can be recovered.

Reserves of silver ore as on 1st April 2010

Mineral	Reserves (million tonnes)	Production (tonnes)	Main producers
Silver	466.985	148.288	Chanderiya Lead and Zinc Smelter, Chittorgarh, Rajasthan, under the control of Hindustan Zinc Ltd., Udaipur; Hutti Gold Mines of Raichur, Karnataka, under the control of The Hutti Gold Mines Co. Ltd., Karnataka.

Source gratefully acknowledged: Statistical Profiles of Minerals 2010-2011 by Indian Bureau of Mines, Nagpur, April 2012.

Figure 8.2: Cyanidation of silver ores followed by its precipitation

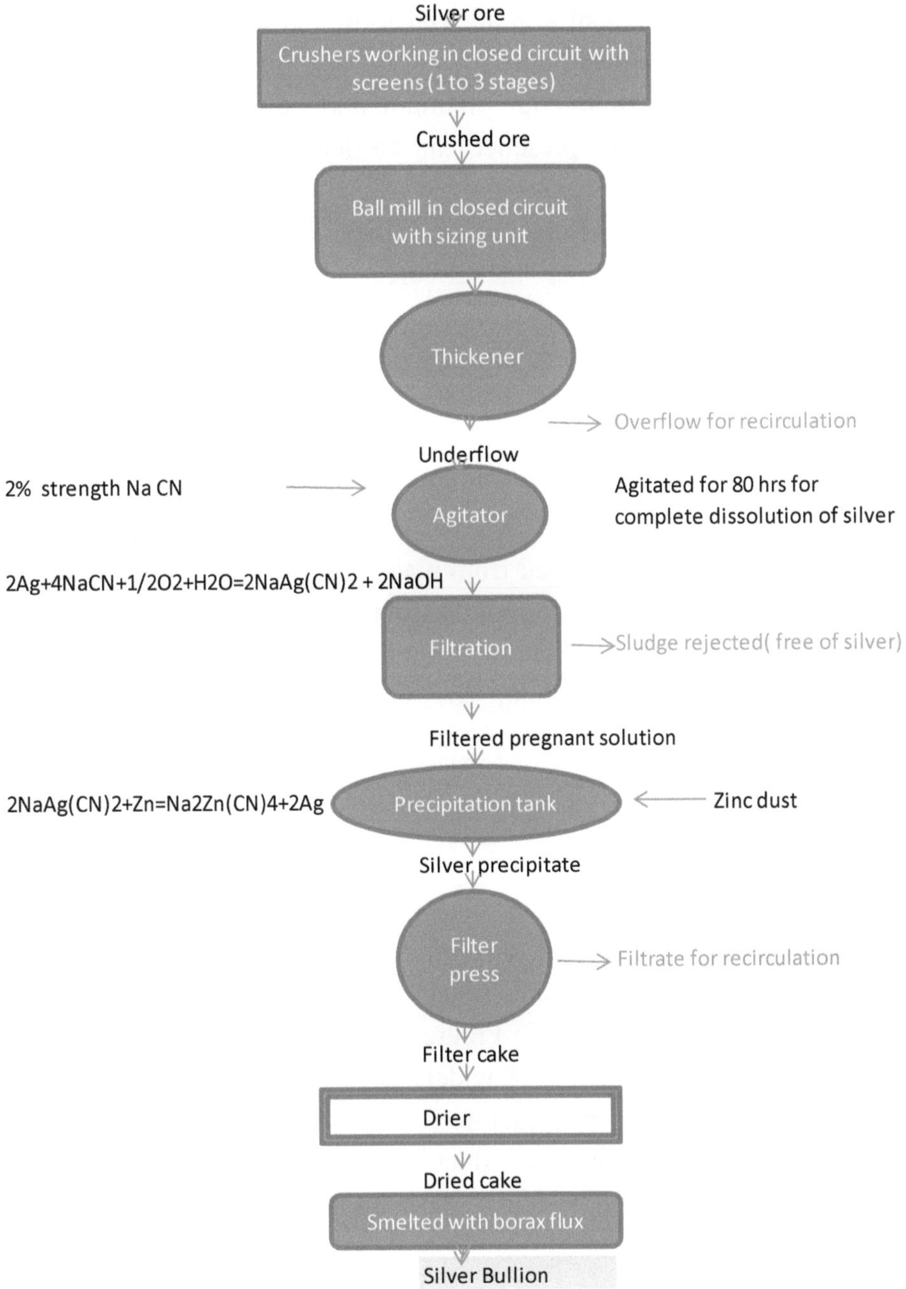

Silver ore

Crushers working in closed circuit with screens (1 to 3 stages)

Crushed ore

Ball mill in closed circuit with sizing unit

Thickener → Overflow for recirculation

Underflow

2% strength Na CN → Agitator — Agitated for 80 hrs for complete dissolution of silver

$2Ag+4NaCN+1/2O_2+H_2O=2NaAg(CN)_2 + 2NaOH$

Filtration → Sludge rejected(free of silver)

Filtered pregnant solution

$2NaAg(CN)_2+Zn=Na_2Zn(CN)_4+2Ag$ — Precipitation tank ← Zinc dust

Silver precipitate

Filter press → Filtrate for recirculation

Filter cake

Drier

Dried cake

Smelted with borax flux

Silver Bullion

8.3 Platinum

Platinum occurs principally in alluvial form and to a smaller extent, as lode deposits (like osmiridium).

Occurrence: Though both platinum and palladium are very rare compared to gold, their presence can be detected widely in traces. In platiniferrous ores, concentration level of platinum is more than palladium. But in copper nickel ores, they are more or less in equal concentrations. In copper nickel ores of Canada, platinum and palladium occur along with rhodium, ruthenium, iridium, and osmium. Platinum is reported to occur in a few places in India such as Tasganan areas, Dangli, and Dangli RF in Lalitpur District in Uttar Pradesh but in trace quantities. Prospecting work is in progress. Platinum and gold are found in metallic form in the alluvial deposits. The tenor of platinum ore is as low as 0.08 ppm.

Platinum can be recovered from placer deposits by gravity methods. Platinum can be separated from the associated gold by using differences in the amalgamation of metals with mercury. Gold amalgamates easily with mercury, whereas platinum does not amalgamate with mercury so easily.

Platinum can be recovered from nickel bearing pyrrhotite by smelting only after separation of nickel. Platinum arsenide is treated by gravity concentration.

Concentration of platinum lode ore (Osmiridium): The ore is usually enriched by crushing, screening followed by gravity methods of concentration producing a concentrate. The platinum metal is recovered by chemical route[1] as shown in Figure 8.3.

Platinum and palladium are widely used for making jewellery. It is also used in dentistry, manufacture of crucibles, dishes and so on. As platinum and palladium are ductile, moderately hard, rare, and corrosion and chemical resistant, both of them are used in making alloys.

---

[1] Dennis,W.H. (1953). *Metallurgy of the nonferrous metals*. London: Pitman.

Reserves of *platinum* :

Mineral	Reserves (million tonnes)	Grade %	Main occurrences
Platinum	19.08	0.00000179 = 1.79 g/tonne	Baula-Nuasahi of Odisha (14.2 MT of 1.7 g/tonne); Hanumalapura in Karnataka (4.5 MT of 1.79 grams/tonne); 0.38 MT at Chettiyampalayan, Tamil Nadu; Mettupalaiyam in Coimbatore and Erode District, Tamil Nadu

Acknowledgement of source: Ajay K. Das of GSI dated 21st May 2012 (from internet).

Figure 8.3: Extraction of platinum from enriched concentrate by chemical route

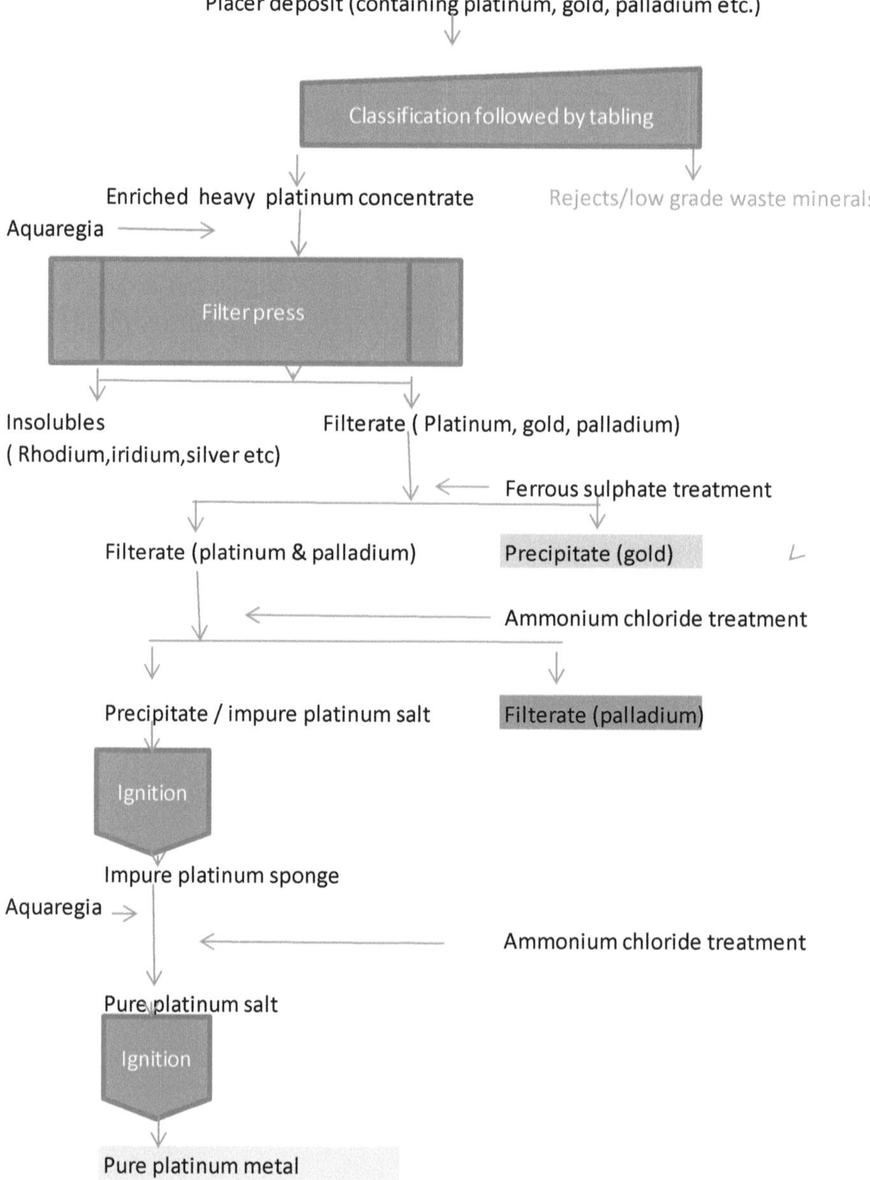

Source gratefully acknowledged: Extraction of precious metals-simplified flowsheet; *Metallurgy of Non Ferrous Metals*, by W.H. Dennis, p. 554.

Chapter 9

Fuels/Energy Sources (Graphite, Coal and Uranium)

9.1 Graphite

Graphite (C) occurs along with silicates and carbonates. Graphite is a readily floatable mineral, even with a frother. Air bubbles get readily attached to graphite, making it easy to float. The crystal structure of graphite makes it ideal for natural flotation. Water-insoluble hydrocarbon oils are used to smear the surface for a better grade and faster flotation. Being soft, graphite smears itself on other undesirable gangue minerals, in turn making them also floatable. This causes difficulty in suppressing unwanted gangue minerals from flotation, resulting in poorer-grade graphite concentrate.

After crushing and coarse grinding, graphite, similar to molybdenum, sulphur, talc, and so on, can easily be floated for obtaining a rougher graphite concentrate and a rejectable tailing. The rougher concentrate is then reground to the requisite liberation size using the flotation reagents such as a soluble frother, water-insoluble hydrocarbon oil at acidic/slightly alkaline circuit with soda ash and water glass. The product is subjected to flotation involving repeated cleanings for achieving the desired purification level of 80% C as shown in Figure 9.1.

Uses of graphite: Flaky graphite is used for making graphite crucibles and electrodes .. For increasing the carbon content of steel, just before tapping, graphite with high carbon and low phosphorus is added.

Reserves, production, and main producers of graphite as on 1st April 2010

Mineral	Reserves (million tonnes)	Production (million tonnes)	Main producers
Graphite	174.850	0.114836	Sivaganga (Tamil Nadu), under the control of Tamil Nadu Minerals Ltd.; Murma, Palamau, (Jharkhand); Khamadih (Odisha), OM&M Pvt. Ltd.; Gandabahali, Naupada (Odisha)

Source gratefully acknowledged: Statistical Profiles of Minerals 2010-2011 by Indian Bureau of Mines, Nagpur, April 2012.

Figure 9.1: Beneficiation of graphite

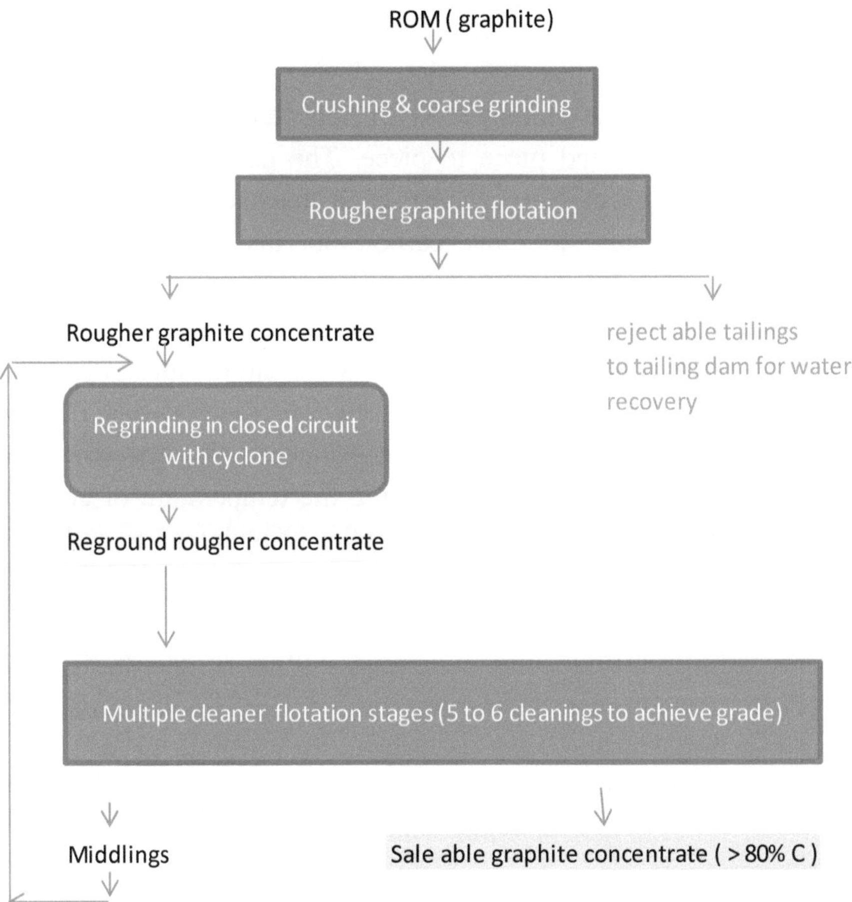

9.2 Coal

Coal is the principal solid fuel. It is highly heterogeneous. Peat contains lot of impurities. Graphite and anthracite are purer coals. Even when extraneous substances like shale, slate, clay, gypsum, and pyrite usually associated with coal are removed, the percentage purity of coal differs from mine to mine and piece to piece. The gangue includes kaolin, other clays, gypsum, calcite, mica, siderite, and so on.

All coals contain not only carbon but also hydrogen, oxygen, nitrogen, silica, sulphur, and carbon to varying extents. They are associated with slate, shale, and so on, besides clay and other minerals. Indian coals have high ash content with low heating values of UHV of 4400 Kcal/kg. It is the cheapest source of primary commercial energy.

Anthraxylon is the coking constituent of coals. The carbonization of coal to obtain coke consists in elevating the temperature of coal. The coking quality of coal depends on the ratios of hydrogen to carbon and oxygen to carbon.

The ash content of the coals differs widely, depending on the presence of associated gangue minerals. Most of the Indian coals contain high ash content. Hence, the ash content has to be reduced to the lowest possible level, through separation/rejection of slate, shale, pyrite, clay, and so on.

Methods of coal beneficiation: The free-ash-forming impurities such as sedimentary partings (bone), bedding planes, slate, shale, clay, can be removed by handpicking and coal beneficiation techniques. These include heavy media separation (HMS), gravity methods (jigging, tabling), and flotation. Beneficiation is required to improve coal quality that meets specifications of clients, for example, iron and steel and power generation industries. Coal washing, as shown in Figure 9.2, involves removal of free/external impurities such as slate, shale, clay, and rock by gravity methods. However, such techniques cannot be utilized to remove fixed impurities, like residual inorganic matter, mineral matter, and a part of pyrites (FeS_2), a major constituent of ash.

Sulphur from pyrites associated with coal can be partly reduced by flotation or HMS. Gravity method is always preferred over flotation because coal floatability gets diminished with its suspension in water. Very fine coals can be floated with cheap reagents such as kerosene and cresylic acid. Pyrite in coal can be depressed with the use of lime.

Principle of cleaning: Mechanical cleaning utilizes the differences in specific gravity of pure coal (say 1.3 to 1.45) from impurities and gangue (specific gravity >2.65).

Floatability of coal: Sink and float tests are carried out to assess the quantity and quality of coal obtainable after coal washing. These involve crushing the coal to a suitable size (say -1/4") and carrying out sink and float tests in liquids having different specific gravities (1.3, 1.4, 1.5, 1.6, etc.). The float fractions at 1.3, 1.4, 1.5, and 1.6 specific gravities, besides sink fraction at specific gravity 1.6 are weighed. The products are analysed for carbon and sulphur content. The test data provides the required information on coal's washability characteristics.

Constrains/limitations with Indian coking coals for steel making: Quality of metallurgical coal available in India is poor, since Indian coals have high ash content, poor coking properties, and are difficult to wash. Hence, steel plants are forced to import low-ash coking coal from countries like Australia and China for blending with available Indian coals having relatively higher ash content.

Reserves, production, quality, and main producers of coal

Fuel	Reserves (million tonnes)	Production/ demand (million tonnes)	Production centres	Main producers
Coal (growth rate: 5.15%)	277,000 MT (7% of world reserves; fourth largest in the world); 100 years life (33,000 MT coking + 244,000 MT noncoking coal)	731 (713.24 MT as estimated by Planning Commission); power sector alone requires 72% of total production	Coal India Limited: Rajmahal of ECL; North Karampura of CCL, Korba, and Hand-Raigarh of SECL, and IB Valley of Talchar of MCL.	Coal India Limited (largest producer in the world) produces 37% of India's requirement. 88% of total coal requirement is met from domestic sources. The remainder 12% is imported from Australia, China, and South Africa

Source gratefully acknowledged: Dr Gautam Dhar's presentation on demand and supply gap and coal security, made at the Associated Chamber of Commerce & Industry on 3 February 2011(from internet).

Figure 9.2: Upgradation of high ash coal by HMS / Chance Cone

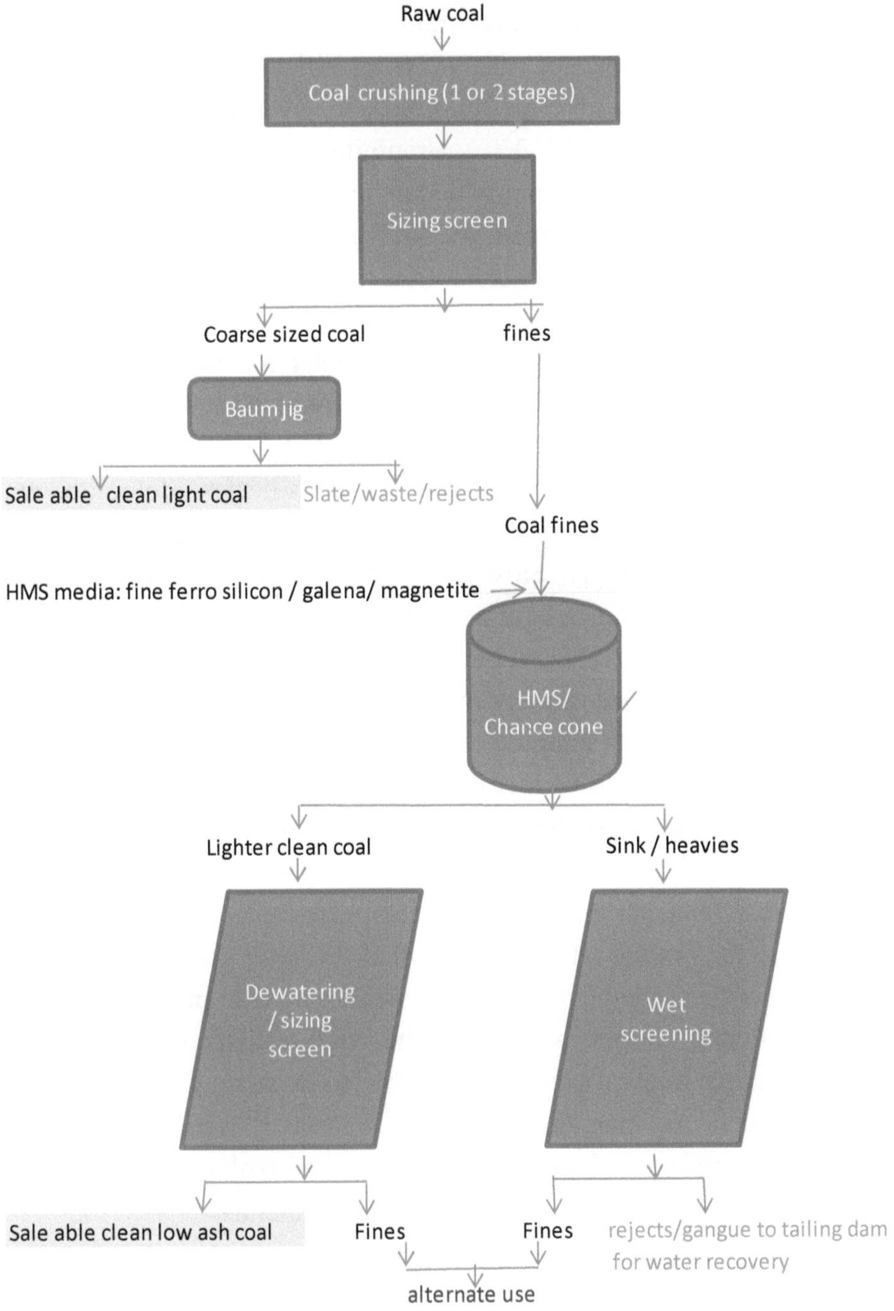

Raw coal

Coal crushing (1 or 2 stages)

Sizing screen

Coarse sized coal

fines

Baum jig

Sale able clean light coal

Slate/waste/rejects

Coal fines

HMS media: fine ferro silicon / galena/ magnetite

HMS/
Chance cone

Lighter clean coal

Sink / heavies

Dewatering
/ sizing
screen

Wet
screening

Sale able clean low ash coal

Fines

Fines

rejects/gangue to tailing dam
for water recovery

alternate use

9.3 Uranium ore

Although it does not occur freely, uranium is widely present at very low concentrations in most of the outcrops/hillocks/deposits. It is as abundant as copper, lead, zinc, tin, gold, etc. The ore-bearing minerals are pitchblende (U_3O_8), Uraninite UO_3UO_2, carnotite $K_2O.2U_2O_3.V_2O_5.3H_2O$. Uranium occurs in small and scattered deposits and can be easily recognized by its yellow or orange colour. Its presence, abundance, and concentration in the ore body are measured by an instrument called the Geiger–Mueller counter.

Since uranium is a precious/strategic metal for country's defence, even a small incidence is adequate for its exploitation. The ore mineral is associated with sulphide minerals and hydrocarbons. Gangue minerals present in the ore include quartz, silicates, dolomite, and calcite. The associated gangue poses difficulty in the separation of ore using flotation technique. However, fatty acids and soaps promote flotation of uranium minerals in neutral circuit.

Separation principle: Uraninite is very heavy mineral (specific gravity: 6.5-8.1). Hence, gravity methods of concentration, as shown in Figure 9.3, are widely adopted. However, as the mineral is soft and goes into slime, it has to be recovered in its coarsest form, as soon as it is liberated in different stages of closed-circuit attrition grinding.

Uraninite and monazite in the Indian beach sands are main sources of uranium. Monazite, the rare earth phosphate mineral (CeLaDy) PO_4, $ThSiO_4$, is the source of thorium with 9% ThO_2 content. The high specific gravity of monazite (4.9-5.3), weak magnetic property, as well as its tendency to get pinned to the high-tension separation roll (nonconducting mineral) make it possible to separate from associated minerals such as magnetite, ilmenite, garnet, zircon, rutile, and hematite. The process of separation of monazite involves the following:

- gravity concentration for separating heavy minerals
- removing its saline coatings by washing to expose their fresh surface and drying
- a combination of high-intensity dry-roll magnetic separators at different strengths

- high-tension separation after pre-heating.

Although flotation of monazite is being used time and again, it has not yet been perfected.

India has been planning to develop nuclear energy as a possible solution for meeting its growing energy needs. As much as 45 tonnes of uranium per year is required for a 220 MV power plant and 100 tons of uranium per year for a 700 MW power plant. As per the policy statement of the Chief Executive of the Nuclear Power Corporation, 14 plants, each capable to produce 700 MW, are likely to be set up during the next five years to cut down power shortage.

Reserves, production, grade of uranium ore, and producing centres

Ore/ mineral	Recoverable reserves of ore (lakh tonnes)	Production centres	Main producers
Uranium ore	17 Tummalapalli (AP) (largest mine in the world); 49,000 tonnes Uranium reserves	Jaduguda, the second mine in Jharkhand (acidic leaching); production is expected to begin soon at the AP plant acidic leaching	Uranium Corporation of India

1. Information source: ‘Nuclear boost: Uranium mine in Andhra could be amongst largest in the world,’ by Abantika Ghosh, TNN, 19 July 2011.
2. As per the published article titled, ‘Thorium reserves’, by Sri Om Prakash Mathur, India has 10.7 million tonnes of monazite containing 963,000 tonnes of ThO_2 (i.e., Indian monazite contain about 9% to 10% ThO_2), from which 846,477 tonnes of thorium can be extracted. The information source is gratefully acknowledged.
3. Thorium, after its conversion to uranium 233 in a nuclear reactor, is proposed to be used for electricity generation. This is in addition to 80,000 tonnes of uranium reserves in India.

Figure 9.3: Beneficiation of Uraninite

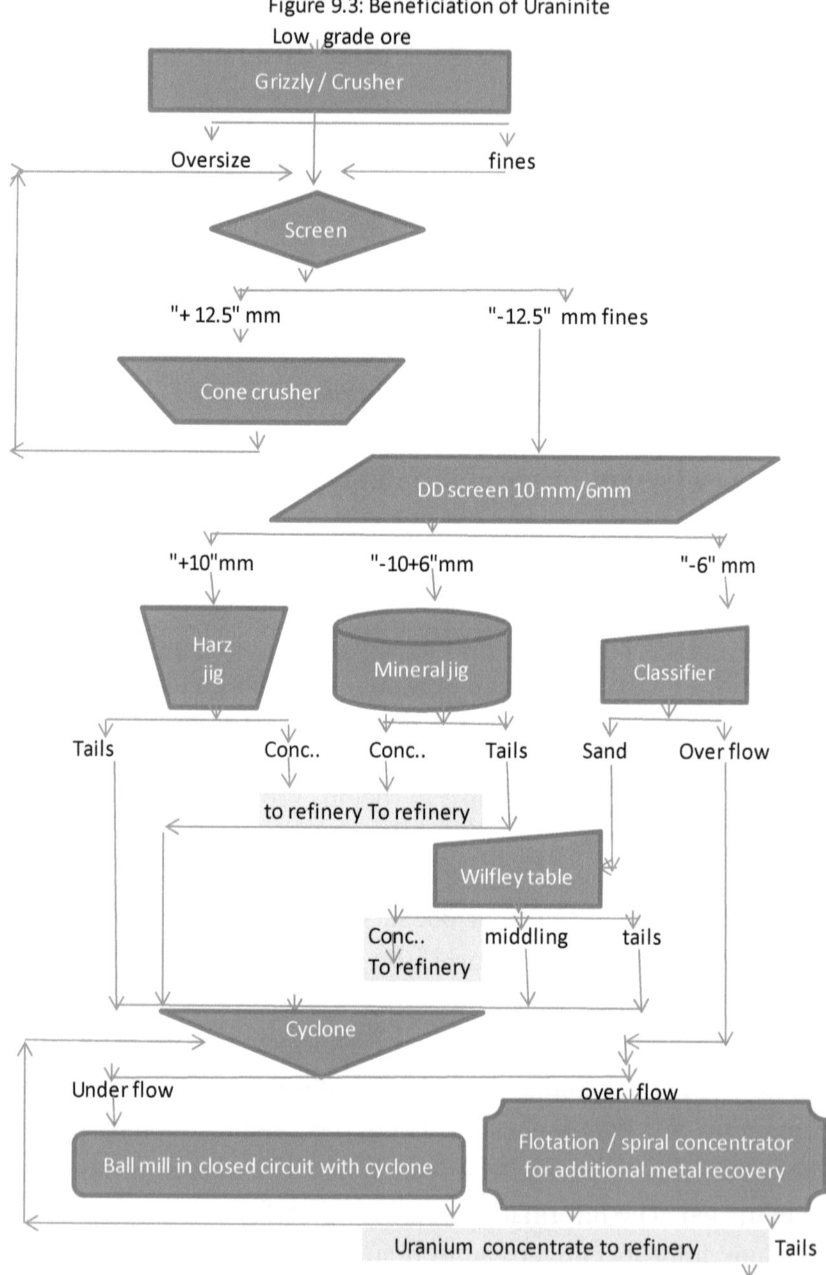

Chapter 10

Beach Sands
(Ilmenite, Monazite, Rutile, Garnet and Zircon)

10.1 Economic minerals' concentration in certain beach sand locations

Beach sand concentrations, available along the east coast of Visakhapatnam, Minavalakurichi, Kerala etc. are important sources of strategic minerals such as ilmenite $FeO.TiO_2$, rutile TiO_2, zircon $ZrSiO_4$, monazite $(CeLaDy)PO_4$, $ThSiO_4$, sillimanite $Al_2O_3.SiO_2$, and quartz SiO_2.

These minerals are more or less liberated/in free state, but perhaps with a saline coating. As much as 70 to 75% of reserves are mostly located in the coastal stretches of peninsular India. The mineral sands are concentrated between the high and low tide lines on the beaches and inland extension of placer sands.

10.2 Separation of the constituent minerals

After removal of saline coatings, these minerals are separated from each other in dry state, utilizing the differences in magnetic properties, electrical conductivities, specific gravities, and surface characteristics as shown in Figure 10.1. For achieving separation, induced roll-type magnetic separators (IRMS) and high-tension separators (HTS) operated at a very high voltage are used followed by flotation and gravity concentration methods. Cleaning is carried out multiple times for meeting purity specifications set by the clients.

10.3 Basis of separation

The following differences in properties of the associated minerals shown in Table 10.1 are utilized for achieving their separation out of beach sands.

Table 10.1: Differences in properties of beach sand minerals

Mineral in beach sand	Specific gravity range	Magnetic property	Response to HTS (thrown from HTS or pinned to HTS)	Mineral floatability
Ilmenite	4.5-5 (heavy)	Strongly magnetic	Thrown (conducting mineral)	
Monazite	4.9-5.3 (heavy)	Weakly magnetic	Pinned (non conducting mineral)	
Garnet	3.2-4.3 (light)	magnetic	Pinned (non conducting mineral)	
Rutile	4.2 (heavy)	Non magnetic	Thrown (conducting mineral)	
Zircon	4.2-4.7 (heavy)	Non magnetic	Pinned(non conducting mineral)	
Quartz	2.7 (light)	Non magnetic	Pinned (non conducting mineral)	floatable
Sillimanite	3.23-3.25 (medium)	Non magnetic	Not amenable	floatable
Magnetite	5.2 (heavy)	Strongly magnetic	Thrown (conducting mineral)	floatable

Following sources of information used in the Table 10.1 are gratefully acknowledged:

Dennis, W.H. (1953). *Metallurgy of the non-ferrous metals*. London: Pitman.

Wills, B.A. & Napier-Munn, T. (2006). *Mineral Processing Technology* **by Barry A Wills, Tim Napier-Munn,** (7th ed.). New York: Elsevier.

Expertise gained from the Consultancy assignment with UNIDO, Jos, Nigeria.

Reserves of economic minerals present in beach sands (sources of occurrence and main producers)

Mineral	**Reserves (million tonnes)**	**Sources of occurrence**	**Main producers**
Ilmenite	278 (largest in the world)	All along the East and West coastal beaches and Kerala	
Monazite	7	Kerala, Tamil Nadu, Seemandhra, Orissa, West Bengal, and Bihar	Indian Rare Earths Limited
Rutile	13		
Zircon	18		
Sillimanite	84	It occurs in Sonapahar of Meghalaya and in Pipra, Madhya Pradesh	
Garnet	86		

1. 'Indian rare earths: Genesis and growth', T.K. Mukherjee, Indian Rare Earths Limited.
2. IREL presentation by Dr R.N. Patra.
   (Both sources of information from internet gratefully acknowledged:

Figure 10.1. Separation of economic minerals from beach sands (containing ilmenite, monazite, rutile, zircon, sillimanite & quartz)

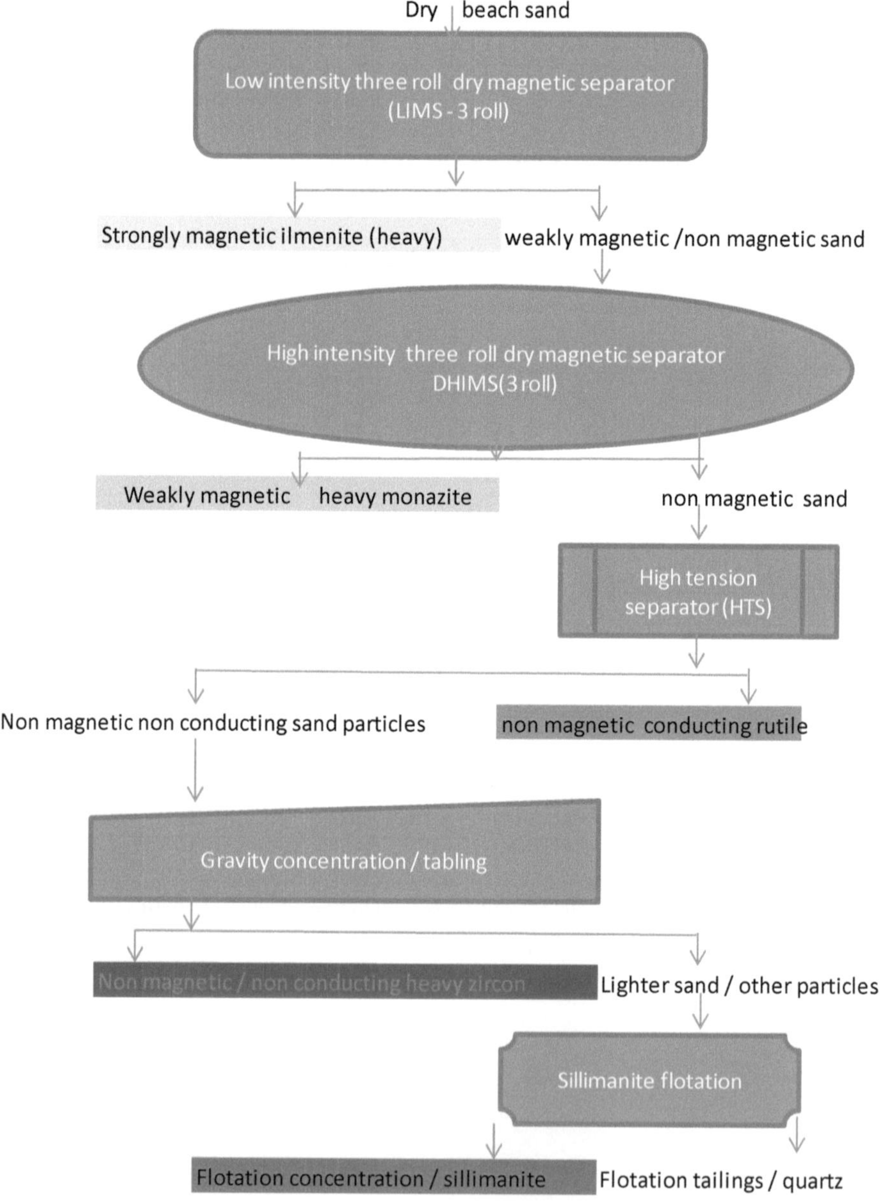

10.4 Mineral composition of beach sand samples

Mineral composition of two representative samples of beach sand from Bhimunipatnam, as evaluated by NMDC Ltd. is given in Table 10.2. The flowsheet for separation of economic minerals from the beach sand sample, based on detailed test work, and as recommended by the R& D Centre of NMDC Limited[11] is given in Figure 10.2.

Table 10.2: Mineralogical composition of two typical samples of beach sand

Mineral	Mineral percentage	
	Sample 1	**Sample 2**
Ilmenite	53.1	83.9
Zircon	4.0	3.6
Monazite	1.0	4.4
Garnet	9.3	5.9
Rutile	1.4	2.2
Quartz and other silicate minerals	30.5	
Magnetite	0.7	
Total	100	100

Information source gratefully acknowledged: Separation of economic minerals from beach sand, at Bhimunipatnam, Seemandhra, R & D Centre, NMDC Ltd.; Hyderabad, January 1992.

[11] National Mineral Development Corporation Limited. (1992). *Separation of economic minerals out of beach sand from Bhimunipatnam*, Hyderabad.

Figure 10.2: Recommended flowsheet for processing Bhimunipatnam beach sand

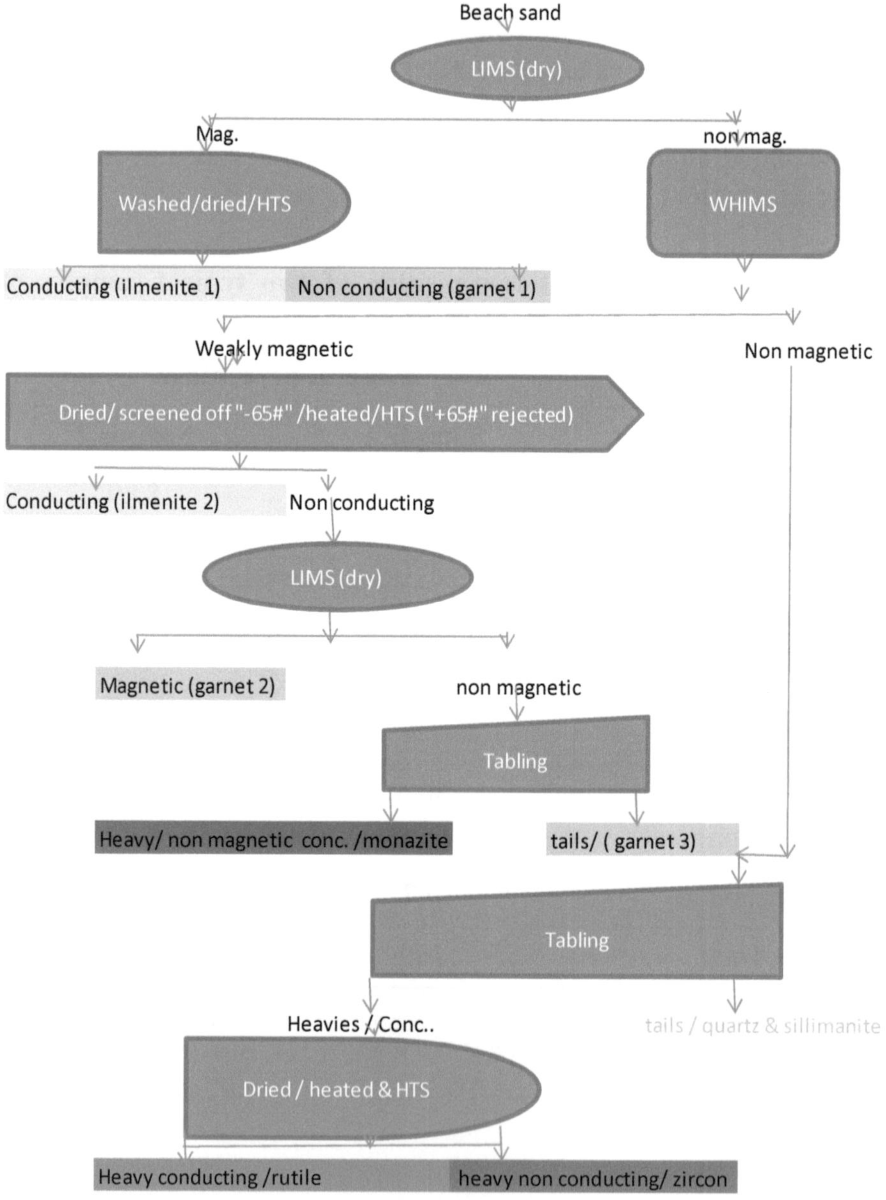

Middlings are always recirculated though not shown in the flow sheet for simplicity and avoid confusion.

Chapter 11

Strategic Minerals

(Tin, Tungsten and Cobalt)

11.1 Tin

Tin ores occur as placer (river sands/sediments) and lode-bearing deposits (outcrops/hillocks). Tin in placer deposits is free, and its concentration is as high as 75% Sn. So tin can easily be recovered from placer deposits by gravity concentration method as shown in Figure 11.1.1.

Lode deposits: Cassiterite ore containing about 1% Sn is associated with one or more minerals like pyrite (FeS_2), chalcopyrite ($CuFeS_2$), Arsenopyrite (FeAsS), gold bearing galena (PbS), sphalerite (ZnS), gold (Au), wolframite $(FeMn)WO_4$, and scheelite ($CaWO_4$). Recovery of cassiterite from the associated minerals present in lode deposits is difficult due to lack of significant differences in specific gravity of the associated minerals. Furthermore, pure cassiterite contains 78.8% Sn. Penalties are imposed for impurities and rewards are given for higher purity. Due to high cost of tin, the tailings are ground and subjected to repeated separation processes for achieving high recoveries of tin and better grades. Hence, cleaning and recleaning operations of the rougher concentrate are carried out to produce higher purity cassiterite of 76 to 77% Sn as shown in Figure 11.1.2. Pure cassiterite is difficult to float. But iron bearing cassiterite can easily be floated from the associated siliceous gangue using fatty acid collectors.

In case gold is associated with tin concentrate, it is passed over a blanket table, and the table concentrate is amalgamated for recovery of gold.

During grinding and classification operations, as cassiterite is friable, it gets over ground and becomes extremely fine and may be lost as slimes in the tailing affecting tin recovery. Thus, in order to produce a clean concentrate with good tin recovery, repeated cleanings of the coarser, interlocked, rougher concentrate as well as processing the tailings in scavenger cells are necessary.

During the processing of cassiterite, the following two fundamental principles of mineral processing have to be strictly adhered to for producing high grade cassiterite concentrate with good recoveries:

1. Mineral has to be recovered in the coarsest size as soon as it is freed from the associated gangue mineral.
2. For economic recovery of any ore mineral from its associated gangue by tabling, closely classified/sized feed is required for good separation by tabling.

Reserves, production, grade, and main producers of tin ore as on 1st April 2010

Mineral	Reserves (million tonnes)	Production (tonnes/ annum)	Main producers
Tin ore Tin concentrate	83.726197 0.102275	61.355	Govindpal, Dantewada of Chhattisgarh Mineral Dev. Corp. Ltd.; Bade Bacheli of Dantewada of Chhattisgarh under the control of Precious Minerals and Smelting Ltd., Jagadalpur, Chhattisgarh

Source gratefully acknowledged: Statistical Profiles of Minerals 2010-2011 by Indian Bureau of Mines, Nagpur, April 2012.

Figure 11.1.1: Recovery of Cassiterite from placer deposits

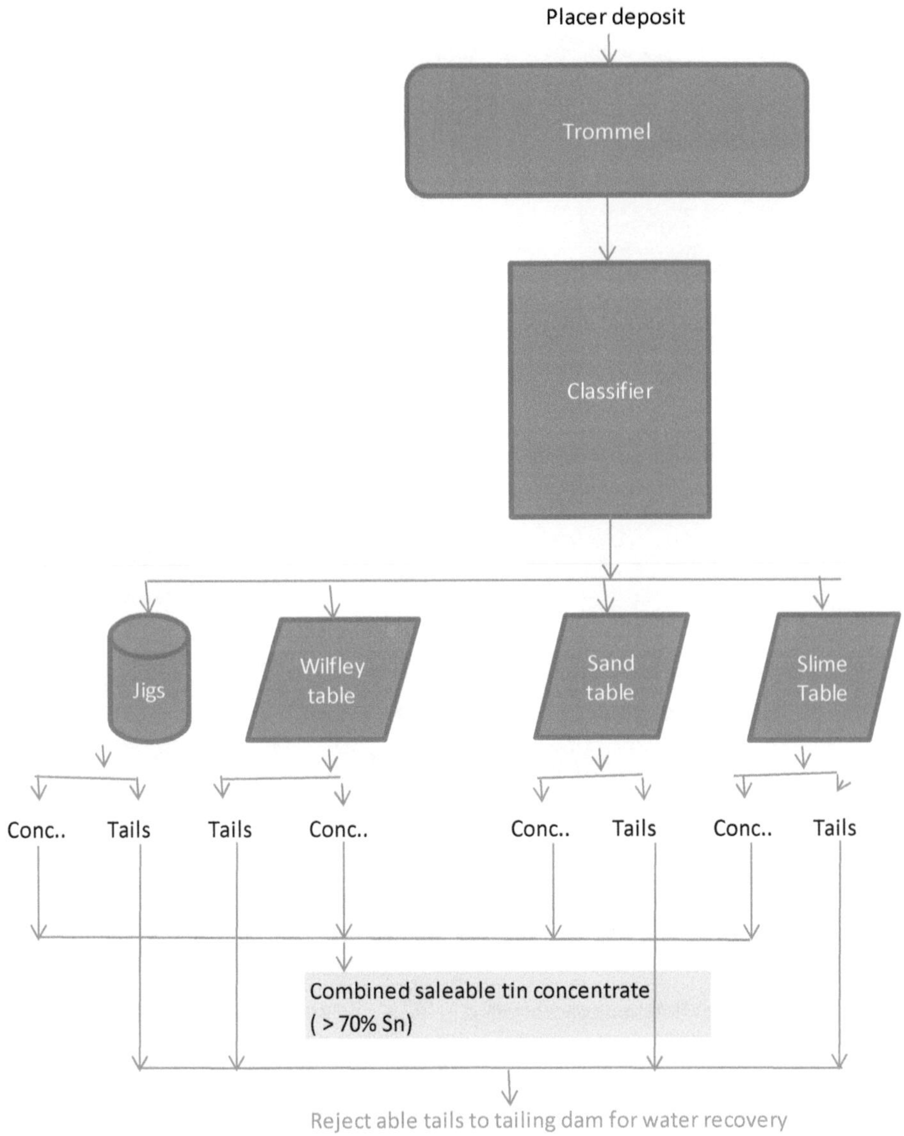

Placer deposit

Trommel

Classifier

Jigs

Wilfley table

Sand table

Slime Table

Conc.. Tails Tails Conc.. Conc.. Tails Conc.. Tails

Combined saleable tin concentrate (> 70% Sn)

Reject able tails to tailing dam for water recovery

Figure 11.1.2: Simple flowsheet for recovery of tin from lode deposits

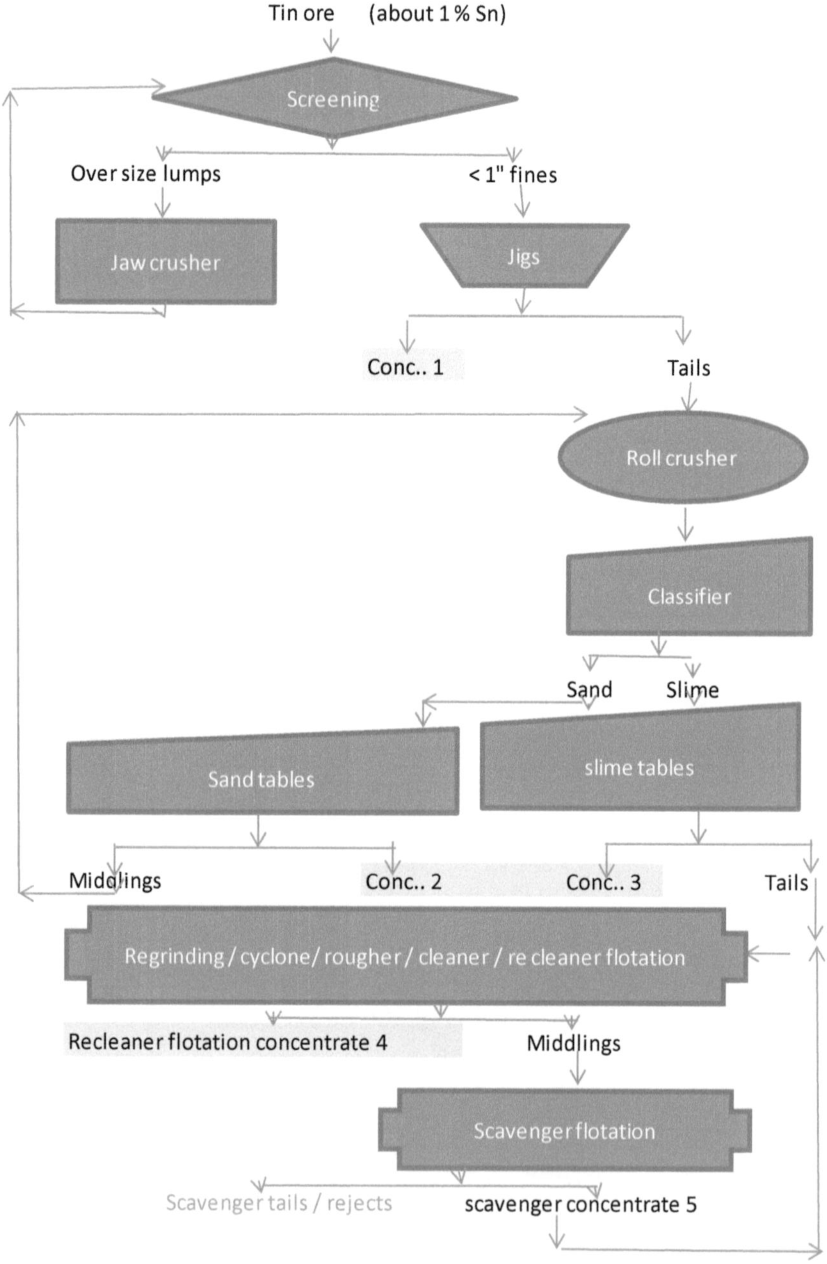

The four streams of tin concentrate are combined & sent to refinery for further enrichment and production of pure metal through removal of impurities.

11.2 Tungsten ore

Wolframite (Fe Mn) WO_4 and scheelite $CaWO_4$ are the two ore minerals. They are associated with cassiterite. Wolframite is magnetic and has high specific gravity (7.2 to 7.5), almost the same specific gravity as cassiterite (6.8 to 7.1). Wolframite is not amenable for flotation.

But both scheelite and cassiterite are non magnetic. Furthermore, scheelite is a fluorescent mineral with blue white colour under UV light. Scheelite is extremely friable. Scheelite is amenable for flotation using oleic acid.

Accordingly, beneficiation of wolframite ore involves sizing, gravity methods, followed by magnetic separation, as shown in Figure 11.2, in order to make use of the differences in hardness, specific gravity, and magnetic property.

Scheelite beneficiation involves coarse crushing followed by sizing for removal of fines. It is necessary to separate scheelite soon after its liberation in the coarsest form possible. Gangue is removed by tabling. Associated sulphides in the table concentrate are floated to produce bulk concentrate of scheelite and sulphide minerals. Pyrite in the bulk sulphide concentrate is roasted to magnetic form followed by its separation from scheelite through magnetic separation. The remaining scheelite concentrate, free from pyrite, is dead roasted to eliminate sulphur and arsenic from non magnetic product.

Tungsten concentrate should contain at least 60% WO_3. Each of the impurities, including tin, copper, phosphorus, arsenic, and sulphur, should be less than 0.5%. Repeated treatment of tailings in successive scavenger cells is necessary to get good recovery of tungsten. In order to make tungsten concentrate meet market specifications, repeated cleanings followed by chemical treatment are necessary.

Tungsten is consumed mostly by steel industries for making alloy steels, die steels, tools, defence equipment, and as filaments in electric bulbs and so on.

The reserves of tungsten ore are estimated to be 83.7 million tonnes. The average grade of the ore is reported to be <0.2%. There is no reported production. Resources are limited. As per IBM's 1992 Report, reserves of tungsten metal were 23,371 tonnes and the reserves of WO_3 were 55,026 tonnes.

Reserves of tungsten ore as on 1st April 2010

Mineral	Reserves (million tonnes)	Production (million tonnes)	WO_3 grade	Main producers
Tungsten ore Tungsten	87.387 0.000142	Not reported	Dagana, Rajasthan: 0.25 to 0.54%; Bankura deposit, West Bengal: 0.1%; KGF tailings: 0.035 to 0.18%; gravel deposits: 0.04%	Govindpal, Dantewada, Chhattisgarh Mineral Dev. Corpn. Ltd.; Bade Bacheli, Dantewada, Chhattisgarh, under the control of Precious Minerals and Smelting Ltd., Jagadalpur, Chhattisgarh

Information Source: Detailed Information of Tungsten ore in India by G.S.I., 1994.

Source gratefully acknowledged: Statistical Profiles of Minerals 2010-2011 by Indian Bureau of Mines, Nagpur, April 2012.

Figure 11.2: Beneficiation of tungsten ore & extraction of tungsten from Cassiterite

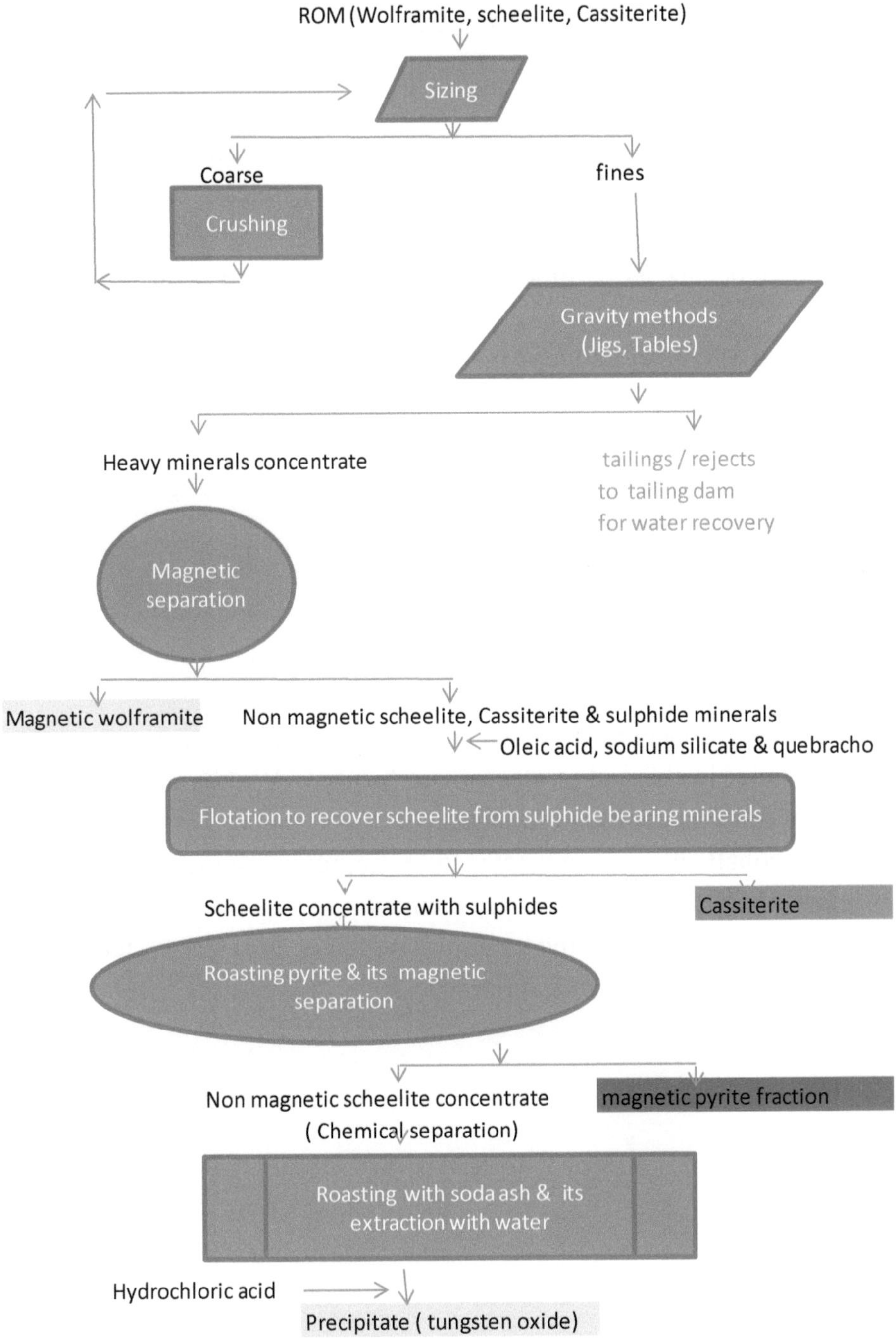

ROM (Wolframite, scheelite, Cassiterite)

Sizing

Coarse

Crushing

fines

Gravity methods (Jigs, Tables)

Heavy minerals concentrate

tailings / rejects to tailing dam for water recovery

Magnetic separation

Magnetic wolframite

Non magnetic scheelite, Cassiterite & sulphide minerals

Oleic acid, sodium silicate & quebracho

Flotation to recover scheelite from sulphide bearing minerals

Scheelite concentrate with sulphides

Cassiterite

Roasting pyrite & its magnetic separation

Non magnetic scheelite concentrate

magnetic pyrite fraction

(Chemical separation)

Roasting with soda ash & its extraction with water

Hydrochloric acid

Precipitate (tungsten oxide)

11.3 Cobalt

Ore minerals: Cobaltite CoAsS, linnaeite Co_3S_4, Carrolite $CuCO_2S_4$ are the ore minerals. The ore may contain about 0.1 to 0.2% cobalt.

Cobaltite ore is processed initially by flotation, followed by a combination of heating in a furnace, chemical treatment, and electrolytic process.

Cobalt occurs as a minor constituent associated with copper and lead ores. Hence, cobalt is recovered as a by-product during the production of copper and lead metals.

Cobalt sulphide is also associated with pyrrhotite. Cobalt sulphide is usually separated by flotation followed by magnetic separation.

In the alternative, pyrrhotite is oxidized by roasting followed by floating off the cobalt mineral.

Brief extraction technique:

1. Selective mining is resorted.
2. Crushing, grinding, and flotation of cobalt sulphide ore producing a flotation concentrate of more than 7% cobalt.
3. The cobalt concentrate is then sintered with lime and coke.
4. The sinter is charged to electric arc furnace. A white alloy of cobalt metal (>40% cobalt) is produced.
5. This cobalt alloy is subjected to refining involving chemical and electrolysis producing the pure element.

Cobalt is used in making ceramics, potteries, blue glass, steel alloys, magnets and for metallurgical purposes.

Chapter 12

Extraction of Aluminium from bauxite

12.1 Ore minerals

Ore minerals include the following: diaspore (Al_2O_3, H_2O), boehmite (Al_2O_3, H_2O), gibbsite ($Al_2O_3.3H_2O$), and corundum Al_2O_3. Bauxite is formed due to weathering of siliceous and argillaceous rocks. Bauxite is associated with hydrated iron oxides and clays with minor amounts of ilmenite. Bauxite, the aluminium ore, is a mixture of trihydrate ($Al_2O_3.3H_2O$) and monohydrate ($Al_2O.H_2O$) with varying amounts of ferric oxide, titanium oxide (rutile or ilmenite), and silica. The alumina content varies from 45 to 65%, silica 1 to 2%, iron 2 to 25% as Fe_2O_3 , combined water 14 to 36%, and up to 2% titanium. However, the composition is not uniform and varies from locality to locality.

12.2 Inherent difficulties in beneficiation

Grinding bauxite produces a lot of slime. So flotation of slimes/ultrafines is difficult. Coarser particles of hydrated aluminium oxides can be floated from the kaolin using soaps, alkaline phosphates, and pyrophosphates. Cationic flotation is not successful.

Corundum Al_2O_3 is valuable if recovered in coarse form. So it cannot be ground and floated. Flotation is therefore not an option for beneficiation of corundum.

For removing excess silica and clay, bauxite washing in log washers or trommels is carried out first. Other methods are not quite effective. Washing is required only to remove clay.

Hence, hydrometallurgical methods are adopted for recovery of aluminium from bauxite/aluminium ores free from clay as per the general flowsheet shown in Figure 12.1.

The trihydrate of alumina is more soluble in caustic soda and, hence, more amenable for beneficiation than the mono hydrate.

$Al_2O_3 + 2\ Na\ OH = Na_2O.Al_2O_3 + H_2O$

(dissolution of alumina at high temperature and pressure).

$Na_2CO_3 + Ca\ (OH)_2 = CaCO_3 + 2\ Na\ OH$

Losses of caustic soda in red mud are compensated with the addition of fresh reagent. This loss is due to the reaction of some caustic soda with silica and alumina forming insoluble sodium aluminium silicate that is discarded along with the red mud. The strength of caustic soda is kept around 10 to 30% through evaporation.

India is a major producer of bauxite and can produce aluminium at internationally competitive prices. About 1 tonne of aluminium is produced from 2 tonnes of alumina (Al_2O_3), which in turn is extracted from 6 tonnes of bauxite.(i.e., 1 tonne aluminium is produced from 6 tonnes of bauxite). Penalties are imposed for high SiO_2 content beyond 5%.

Major bauxite reserves are concentrated along the east coast of Orissa and Seemandhra.

Reserves, production, grade, and producers of bauxite as on 1st April 2010

Total production of aluminium: 2.1 LTPY

Mineral	Reserves (million tonnes)	Production (million tonnes)	Principal states of bauxite occurrence	Major production centres
Bauxite (amounting to 7.5% of world reserves- fifth largest reserves) About 89% reserves of 40 to 50% Al_2O_3 are of metallurgical grade	3479.62	12.641	Orissa, Chhattisgarh, Seemandhra, Madhya Pradesh, Gujarat, Maharashtra, Karnataka, and Bihar	NALCO, Damonjodi, (Odisha) ; and Chennai, (Tamil Nadu); BALCO, Korba, (Madhya Pradesh); INDAL Muri (Bihar), and Belgaum, (Karnataka); HINDLCO, Renukoot, (Uttar Pradesh)

Source gratefully acknowledged: Statistical Profiles of Minerals 2010-2011 by Indian Bureau of Mines, Nagpur, April 2012.

Figure 12.1: Extraction of aluminium from bauxite by Bayer's process

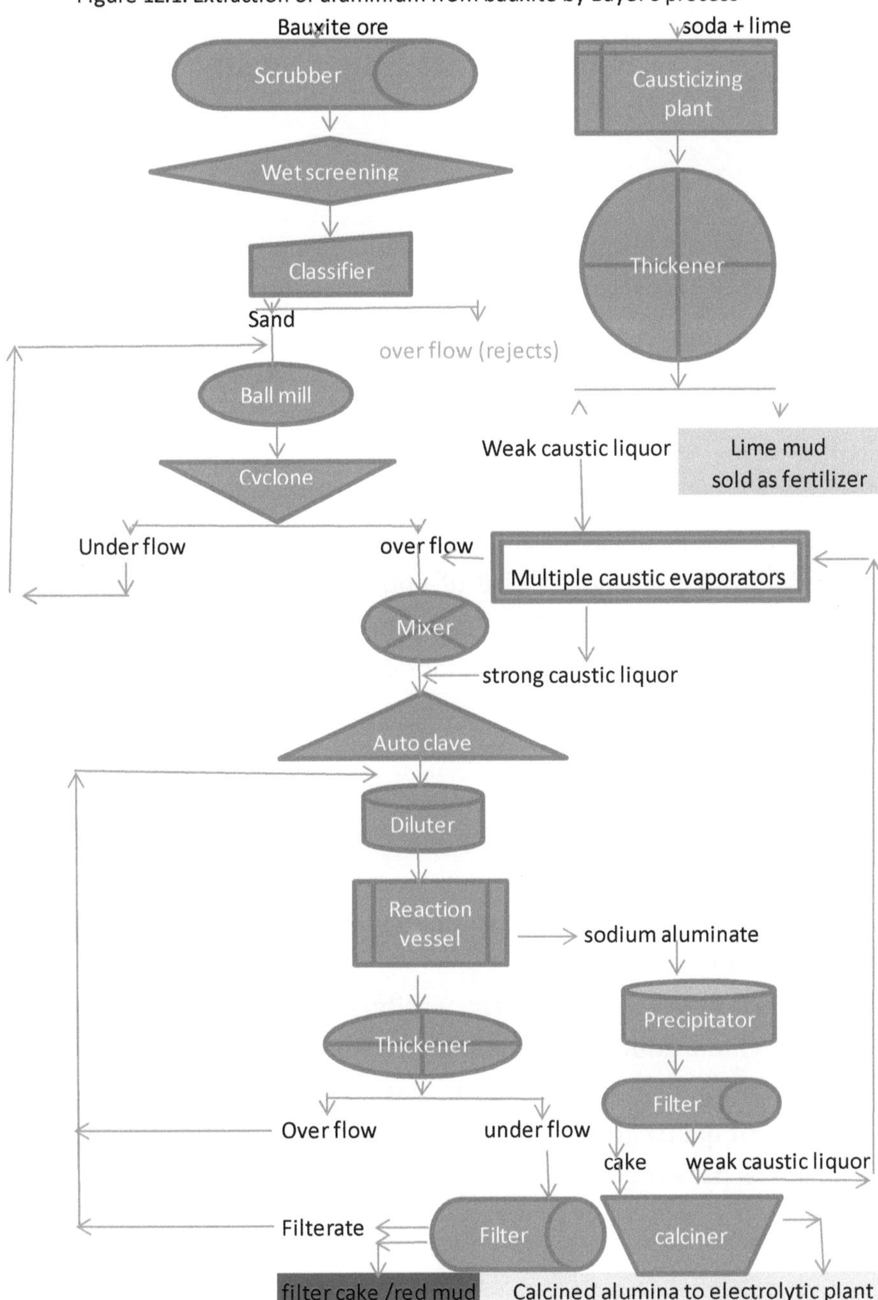

Alumina production: - (Courtesy:Metallurgy of Non ferrousMetals by W.H.Dennis)

Chapter 13

Quartz and Silica Sand

13.1 Quartz

Quartz (SiO_2) is the most widely available gangue mineral associated with most of the ore minerals. In the purest form, it is not floatable by soaps, fatty acids, or their salts. However, it gets activated by many salts of calcium, barium, copper, lead, zinc, iron, and aluminium. In the absence of an activator like barium, quartz hardly adsorbs laurate collector and is therefore unable to float by itself. Nevertheless, in the presence of an activator like barium ions, quartz adsorbs both activator and more laurate collector at higher pH and readily floats.

Deactivator of quartz: Use of common de-activators such as soda ash (Na_2CO_3), water glass (sodium silicate Na_2SiO_3), or alkali phosphate prevent excessive flotation of quartz. If the activating ion responsible for flotation of quartz is not copper, other suitable deactivating regents have to be used.

Flotation of quartz: For flotation of quartz with soaps and anionic collectors, it is necessary to keep the activating ion concentration lower than the appropriate critical level of concentration at a given pH. At higher concentrations of activator ion above the critical level, quartz would not float.

Amine flotation: Quartz can easily be floated even with a mono layer of amine collector in the alkaline range of 10.5 to 11 pH. Hence, amine dosages would be quite low (cationic flotation). Cationic flotation of quartz from low grade iron ores like BHQ/BHJ is generally adopted as per the general flowsheet shown in Figure 13.1.

Reserves, production, grade, and main producers of quartz as on 1st April 2010

Mineral	Reserves (million tonnes)	Produc-tion (million tonnes)	Principal states of occurrence	Major production centres
Pure quartz	3499	0.456829	Seemandhra, Rajasthan, Jharkhand, Gujarat, West Bengal, Maharashtra	Kabra, (Rajasthan); Warlaparthi and Balanagar of (Seemandhra); Karma, (Jharkhand)

Source gratefully acknowledged: Statistical Profiles of Minerals 2010-2011 by Indian Bureau of Mines, Nagpur, April 2012.

Figure 13.1: Recovery of quartz from low grade iron ore

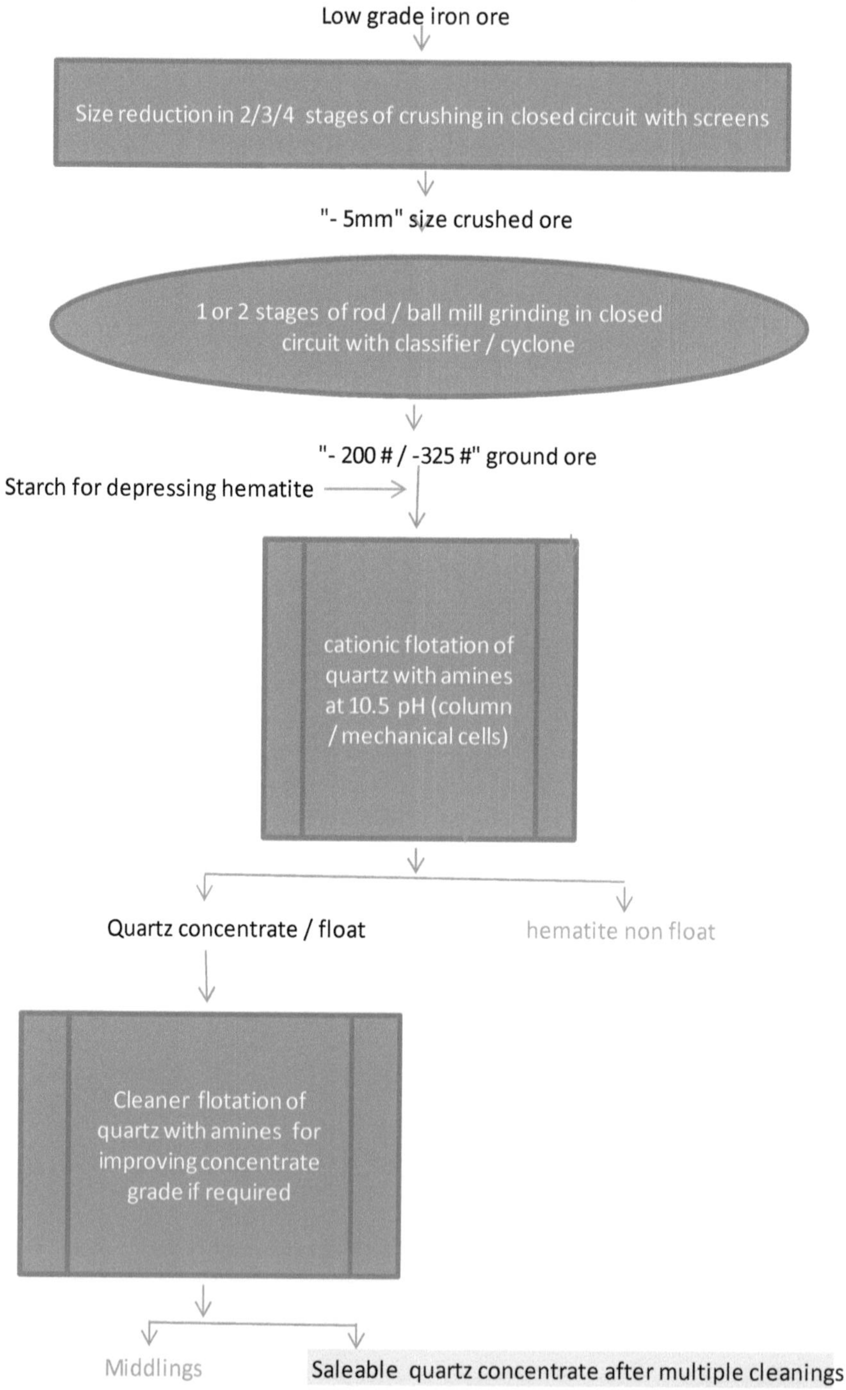

Low grade iron ore

↓

Size reduction in 2/3/4 stages of crushing in closed circuit with screens

↓

"- 5mm" size crushed ore

↓

1 or 2 stages of rod / ball mill grinding in closed circuit with classifier / cyclone

↓

"- 200 # / -325 #" ground ore

Starch for depressing hematite →

↓

cationic flotation of quartz with amines at 10.5 pH (column / mechanical cells)

↓

Quartz concentrate / float | hematite non float

↓

Cleaner flotation of quartz with amines for improving concentrate grade if required

↓

Middlings | Saleable quartz concentrate after multiple cleanings

13.2 Silica sand

For its use in glass industry, the sand should be free from iron bearing heavy minerals, clay, silt, stains, and so on. Processing techniques of silica sand vary depending on its associated impurities. Pure silica sand is required by several industries, including steel making, foundries, making concrete, and so on. Generally, the pure quartz rock is selectively mined, crushed in jaw/cone crushers, ground in Raymond Mills/grinding rolls to finer sizes, followed by screening in large capacity Derrick screens/Rotex screens to meet customer specifications. Screens having high screening capacity of 250 TPH are commercially used. Further processing of silica sand depends on associated impurities.

Just washing is adequate to produce pure sand that is free from iron-bearing heavy minerals, clay, silt etc. This is the cheapest method. If the sand is associated with clay, silt, or stains, washing the sand followed by hydro cycloning is needed. While pumping the wet sand through pipes to the hydrocyclone, inherent attrition of sand grains helps in removal of clay, silt, and stains associated with sand. In case the iron bearing heavy minerals or mica are associated as impurities with sand, anionic flotation is adopted to float off the gangue.

The unit operations are slightly modified so as to meet the product specifications of different user industries. The industry prefers dry sand, high in SiO_2 content (>90%), and specified size range. Several processing plants producing silica sand as per the flowsheet shown in Figure 13.2 are in operation in South India.

The final wet sand meeting market demand/specifications is subjected to costly drying operation, just to help the user industry like glass makers for improving flow ability characteristics of silica sand. In turn, smoothly flowing dry sand is easy to measure/weigh correctly and achieving thorough mixing it with other ingredients before subsequent production steps. This facilitates production of saleable end product and adhere to the high standards set by the user industry.

Reserves, production, grade, and main producers of silica sand and quartz as on 1st April 2010

Mineral	Reserves (million tonnes)	Production (million tonnes)	Principal states of occurrence	Major production centres
Silica sand and quartz	3,499	3.081	Seemandhra, Gujarat, Maharashtra, Rajasthan, Uttar Pradesh	Amod, Motibar, Bhavani of Gujarat; Momidi and Chinthavaram of Seemandhra

Source gratefully acknowledged: Statistical Profiles of Minerals 2010-2011 by Indian Bureau of Mines, Nagpur, April 2012.

Figure 13.2: Production of silica sand from pure quartz veins

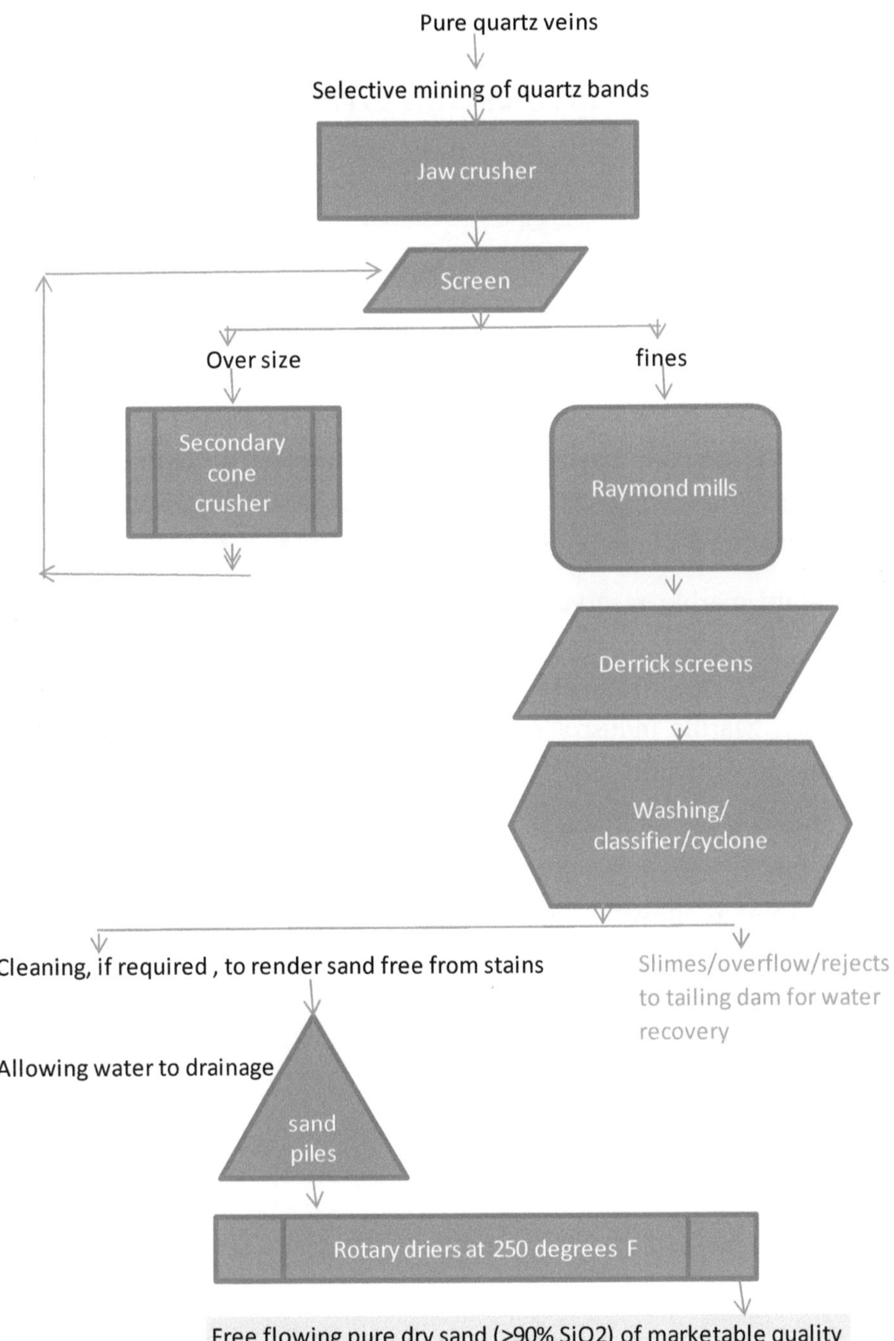

Chapter 14

Dimensional Stones

(Marbles and Granites)

14.1 Marbles

Rajasthan has major deposits of marble and accounts for 90 to 95% of India's production. The state contributes 50% of minor mineral production in India. Production of Makrana marble having 90 to 98% $CaCO_3$ is up to 1.2 lakh tonnes per year. Among other available varieties, "Ambaji marble" is the most preferred marble in India. Producers like Dungri M, Dungri Makrana, and Dungri Marble India have developed a variety of marbles.

White marble reserves are estimated to be 1,200 million tonnes. Marble processing is perfected by trial and error. Flora Marble India, Mumbai, is one of the big factories in India with high production capacity. Marble industry of India is the fourth largest mineral-based industry in the world.

The process involves mining big marble slabs, cutting them into requisite sizes through hacksaw machines, polishing them prior to their sales/export as shown in Figure 14.1.

Figure 14.1 : Processing marble stones

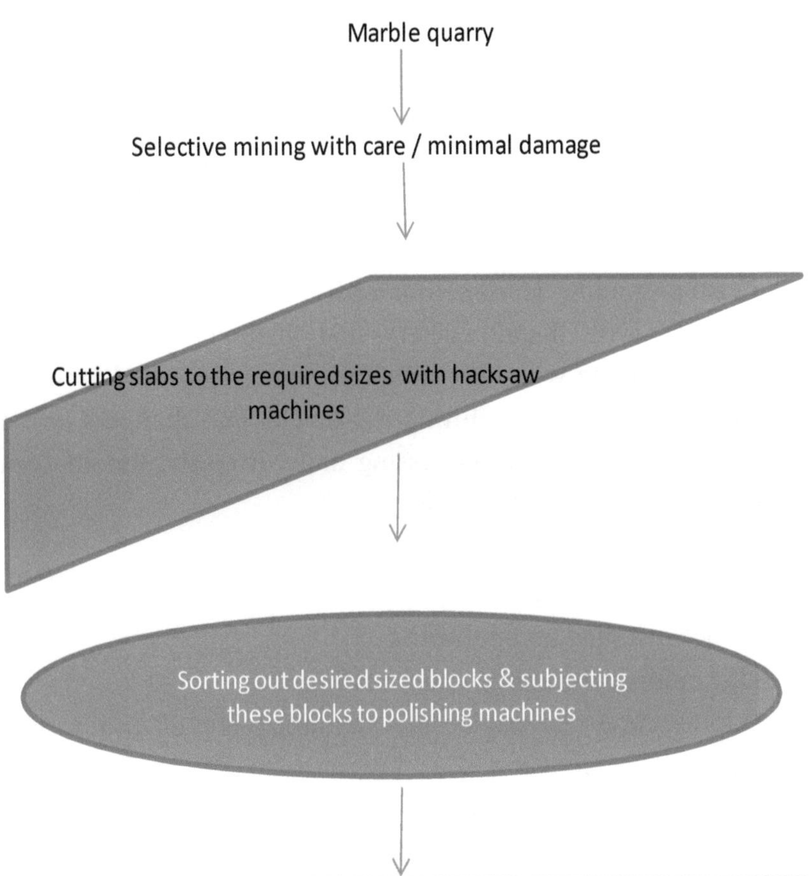

Exporting cut , sized and polished marble slabs through trucks / goods trains/ shipping

14.2 Granites

Granite is a very hard crystalline igneous or metamorphic rock containing mostly feldspar and quartz.

India has the largest reserves of granite in the world. The southern and eastern belts have abundant granite deposits. Granites are the most sought after material in construction of buildings, monuments, massive structural works, airports, flooring, tiles and so on. Highly polishable Indian granite (with mirror polish) is well known in the international market for its elegance, aesthetic quality, durability, and scratch-free glossy surface.

Granite mining is done manually, using drilling machines and channelling with hand chisels and hammers. Large mines use compressors along with the drilling machines, followed by blasting. All mines use big cranes for lifting the blocks into dumpers or trucks for transporting them to the processing units. Broadly, the unit operations as shown in Figure 14.2 include:

- Dressing
- Cutting and sawing
- Surface grinding and polishing
- Edge cutting, trimming, and rounding.

Due to these operations, the yield/recovery of granite is low but profitable. The recovery is as low as 6 to 7% in the case of black granites (most preferred variety) and 10 to 12% in the case of pink and white granites. The rest is waste. Still smaller size granite slabs and granite tiles are prepared out of the waste, after removing the saleable granite blocks, for use as kitchen/bathroom tiles. The waste is usually associated with silicates, iron oxides, fluorite, and calcite.

There are about 20 big companies, each producing and exporting about seven container ships of granite per month. About 85 to 95% of our production caters to the export market generating a revenue of more than Rs.4,000 crore. Rajasthan alone contributes about 95% of export market for dimensional stones. Still, as per the claims of industry, the export market is not so lucrative due to high royalty, export tax, railway freight, port and shipping charges, and so on. No doubt the Government earns quite a lot of revenue out of granite mining.

There are about 100 main companies located in Tamil Nadu, Karnataka, Seemandhra, Maharashtra, Rajasthan, and so on producing granites valued at US $ 85 million/annum. The leading companies are expanding their mines and processing units using highly sophisticated machinery to meet the growing domestic/export demand. Granite industry is growing by 50% annually.

Reserves: As much as 46,230 million cubic metres of granite is estimated to be available as reserve.

Environmental concerns: The granite industry has serious social and environmental concerns. As the available water is drawn/used up by the granite/marble industry in Rajasthan, the local farmers are deprived of water for farming/cultivation. So, for livelihood, the farmers having their own agricultural lands are forced to become labourers in the granite industry located in their own villages/neighbourhood.

Figure 14.2: Processing granites

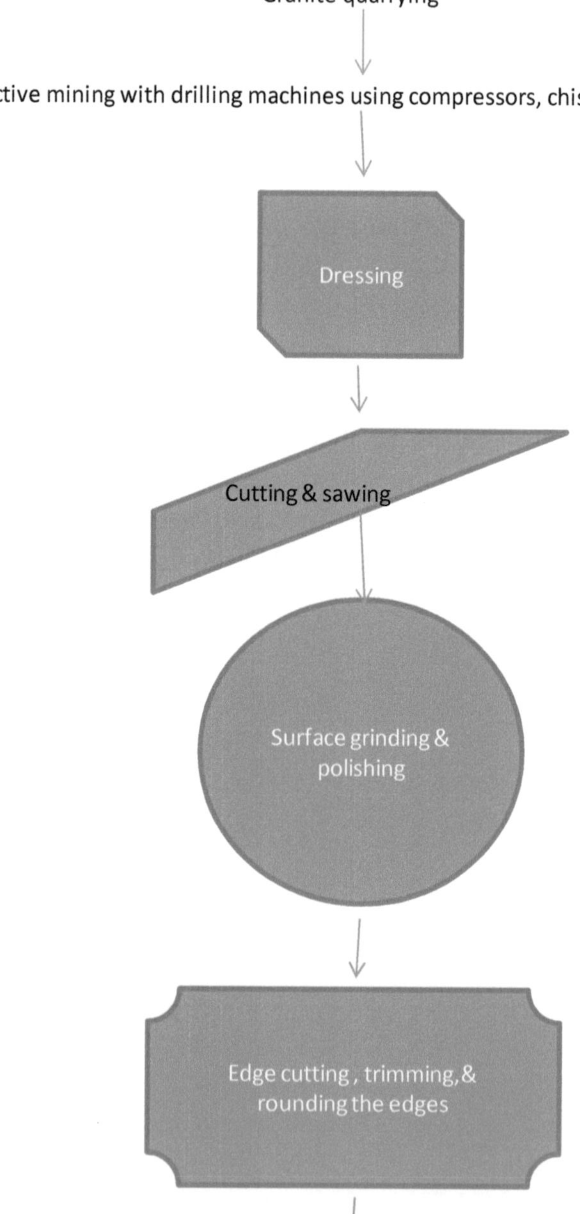

Chapter 15

Preparation of Flowsheets (Process, Material Balance and Equipment Flowsheet)

15.1 Preparation of process flowsheet

Tests have to be carried out with a representative sample of the ore, representing the deposit, in order to evolve an optimum process route to achieve marketable quality of the final concentrate with desired physical, chemical, and metallurgical properties as well as achieve maximum possible yield and metal recovery. This process flowsheet indicates the optimum route through which pre-specified and reproducible test results are obtainable under a given set of operating conditions.

For illustration purposes, test results obtained by the R&D Centre of NMDC Limited, Hyderabad, with a sample of Siri Goa Iron Ore, Goa, undertaken at the instance of Chougule & Co. Ltd., are used as a basis for the preparation of process, material balance, and equipment flowsheet. The plant is designed to process 300 TPH ROM. Figure 15.1 shows the basic process flowsheet for achieving the desired quality of the final concentrate under specified operating conditions.

As this is meant to be only an example, only major unit operations are mentioned:

Figure 15.1: Process flowsheet for Goa Iron ore

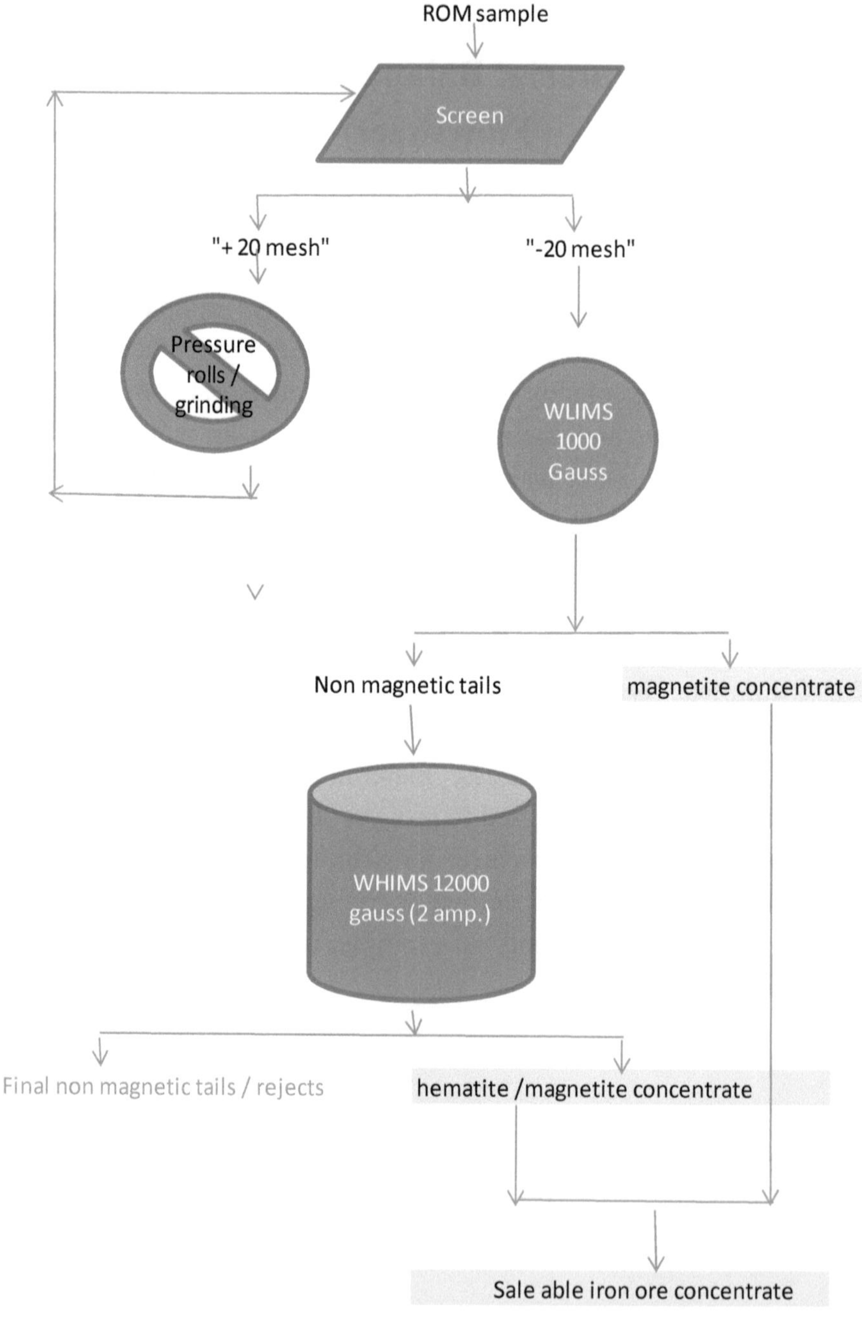

15.2 Material balance flowsheet

The material balance flowsheet as shown in Figure 15.2 provides information on loads to different equipment, product weight percentages, water consumption (total and net); flow rate of solids, water, and pulp, product quality, and specific gravity of pulp. For the purpose of illustration, a simplified material balance flowsheet with weight percentages and quality of the products only is presented in Figure 15.2.

Figure 15.2: Material balance flowsheet

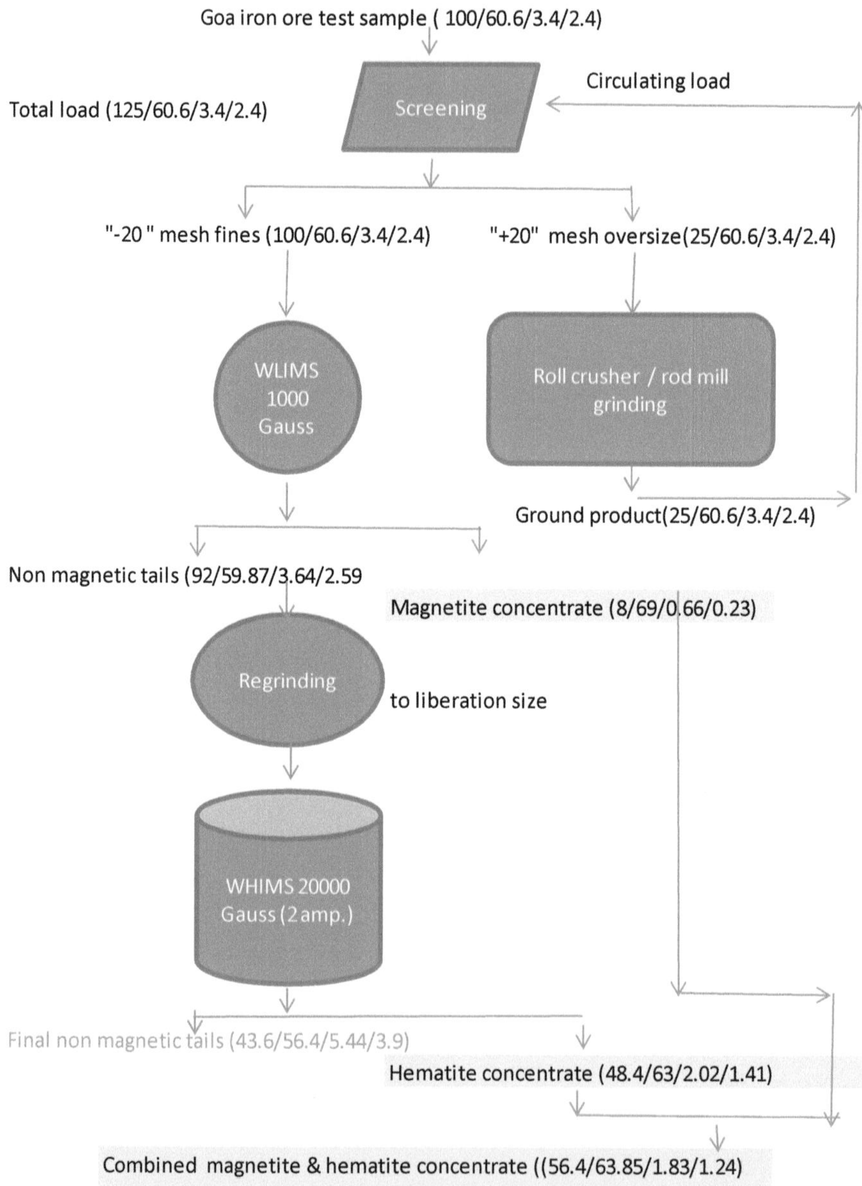

15.3 Drawing up specifications

Equipment specifications are prepared based on the loads to different equipment, which in turn can be known from the material balance flowsheet as shown in Table 15.2. Based on these specifications, bids are invited for supply of equipment meeting the specified duty requirements of 300 TPH ROM. As this is by way of illustration, only a few equipment are included in Table 15.1.

Table 15.1: Drawing up equipment specifications

Sl. No.	Equip-ment name	TPH	Size/ source/ model no.	Quantity required	KW approx. (each unit)	Settings
1	Ore bin	600 tonnes	300 M^3	1	-	1 m × 1 m opening
2	Grizzly feeder	300	HGF 1248	1	5	20 mm opening
3	Double deck vibrating screen	300	6'W × 16'L	2	10 × 2 = 20	6 mm/1mm
4	Pressure rolls/rod mills	300	McNally Bharat	2	50 × 2 = 100	Product: 20 mesh
5	Low intensity wet magnetic separator (LIW MS)	350 (30% solids)	600 mm W × 2400 mm L; perma-nent magnetic type (rare earth), Eriez/ Boxmag	2	2 × 10 = 20	Concentrate grade: 69% Fe; $SiO_2 + Al_2O_3$: <2.5%
6	Water pump	200 m^3/hr; Head: 20M	McNally Bharat	4 + 1 standby	40 × 4 = 160	Water contains <80 ppm solids
7	Wet high Intensity magnetic	300 = 4 × 75 TPH each	Hum-boldt Wedag,	4	100 × 4 = 400	8R grooved plate, max. 20000 Gauss,

	separator (WHI MS)		Gauss: 12,000; amp: 2.5			2.5 mm gap, 35% solids
8	Slurry pumps (Warman / McNally Bharat)	Rubber/ ceramic-lined 200 m^3/hr corrosive solids	McNally Bharat/ Warman	4 + 1 standby	50 × 4=200	Anticorrosive lining/ 40-50% solids pulp; 10m head
9	Concentrate thickener with overload alarm, auto rake lifting device, etc.	180 tonnes solids; 35% feed; underflow discharge: >.50%	Hindustan Dorr Oliver/ EIMCO KCP	1	10 × 1 = 10	Underflow: >50% solids, overflow: < 80 ppm solids
10	Filter with vac. Pump, compressor, water pump, moisture trap etc.	180	Hindustan Dorr Oliver/ EIMCO KCP	1	105 × 1 = 105	Filter cake: <9.5% moisture, clear filtrate
11	Belt conveyors 36” wide	400		10 Nos.	10 × 5 = 50	
12	crane/ workshop equipment/ welding machine/ Misc. etc.			One set	200	
	Total load KW				1270	

Optimum operating conditions.

Pressure rolls/grinding in close circuit with a 20 mesh screen at 60% solids.

Wet low intensity magnetic separator: 1000 Gauss at 30% solids

Wet high intensity magnetic separator (power input: 2 amp, Gauss 12000, 40-50% solids at 2.5mm gap, Matrix: 8R grooved plate).

Note: To make it brief, only essential equipment are shown in the table.

This is based on the knowledge gained from the UNIDO Consultancy assignment at Jos, Nigeria. Thanks are due to UNIDO, United Nations, and Federal Government of Nigeria.

15.4 Equipment flowsheet

Figure 15.3 illustrates the number of equipment required for processing 300 TPH ROM. The plant layout, not included in this book, shows the special arrangement of the equipment to be installed in different floors and their interconnection with conveyors, slurry pumps, feeders, bins, chutes, and so on for smooth, continuous operation of the plant.

Figure 15.3 : Equipment flowsheet

S. No	Equipment name	Quantity	Settings	Symbol
1	Ore bin	1	3000mm opening	
2	Vibrating grizzly	1	10 mm openings	
3	vibrating screen	2	30mm/6 mm	
4	Pressure rolls/ grinding mills	2	"- 100" mesh	
5	Low intensity wet magnetic separator	2	1000 gauss	
6	Wet high intensity magnetic separator	4	12000 gauss, 2 amp	
7	Thickener	1	1 rpm UF: > 50% Solids	
8	Filter	1	Cake:<9.5 % moisture	

To render the illustration information less complex,
other equipment such as belt conveyors, slurry pumps, water pumps, and other process control equipment, instrumentation, though required, are not shown.

Table 15.2: Specifications of required equipment

Equipment no.	Name of the equipment	Qty.	Throughput TPH	Size	KW	Settings
1	Ore bin	1	600	300 M^3	-	1 m × 1 m opening
2	Vibrating grizzly	1	300	HGF 1248	1 × 5 = 5	20 mm opening
3	Double deck vibrating screens	2	300	6' × 16'	2 × 10 = 20	6 mm, and 1 mm opening
4	Pressure rolls/ grinding mills	2	300	To meet duty requirement	2 × 50 = 100	Product: <20 mesh
5	Low intensity magnetic separator	2	175 × 2 = 350	600 mm W × 2400 mm L	2 × 10 = 20	1000 gauss; 69% Fe concentrate
6	Wet high intensity magnetic separators	4	75 × 4 = 300	To meet duty requirement	4 × 100 = 400	12000 gauss; +63% Fe concentrate
7	Concentrate thickener	1	180	To meet duty requirement	1 × 10 = 10	>50% solids in underflow
8	Concentrate filter	1	180	To meet duty requirement	1 ×105 = 105	< 9.5% moisture
9 to 12	Other equipment				610	
	Total load (KW)		300		1270	

Chapter 16

Setting up a Beneficiation Plant

16.1 Major milestones to be achieved in setting up a commercial beneficiation plant

Following are some of the major steps to be taken up prior to setting up a commercial beneficiation plant:

- Estimation of the total proved ore reserves through drilling, ore to waste ratio, extent and depth of the ore body, and determination of the grade and mineralogy of the ore.
- Getting the ore tested both on a bench/laboratory as well as, if possible, continuous pilot plant scale to know the quality and quantity of marketable concentrate; metal/element recovery of the desired constituent; and likely variations in the quality of the concentrate, including the extent of objectionable impurities like phosphorus, sulphur, and so on likely to be present in the saleable concentrate.
- Getting statutory clearances, prospecting/mining lease, Government approvals, forest and environmental clearances, and approvals of mine plan and production schedules.
- Preparation of TEFR/DPR by the organization/competent agencies including the optimum scale of production, capital and production costs, profitability over the prevailing/fluctuating spot market rate/ auction price/export price, Rupee exchange value, and so on.
- In case the project is commercially viable and profitable, organization should prepare financial plan, seek bank loans, and readying equity by the promoters.
- Arriving at the optimum process route, determining material balance and preparing equipment flowsheet and tentative plant layout for production of required quantity of the marketable concentrate while keeping scope for expansion.

- Acquisition of the required mining lease (ML) for the ore deposit and land for locating processing plant, waste dumps, water reservoir, concentrate storage yard, tailing dam/disposal ponds, staff quarters/amenities, generator, railway siding if justified, and so on.
- Arranging/seeking required power and water, recruiting competent manpower in phases as per the needs, and training them to discharge their duties efficiently.
- Taking investment decision based on availability of capital, profitability, and market demand.
- Invitation of equipment bids, selection, ordering equipment, and their receipt.
- Fixing agencies for the execution of the Project on a time bound basis to avoid cost overruns on account of time delays in project completion.
- Execution of Project, including equipment inspection, finalization of equipment layout, fabrication, erection, commissioning, finetuning plant operations/fabrication/adjustment of chutes and machinery for smooth flow of solids, and rectifying bottle necks if any.
- After the plant is set up, undertaking performance guarantee tests to ensure production of the final product as per the projected quality, and quantity as per the DPR, after fine tuning, if necessary.
- Only by way of example, some general details of performance guarantee test conducted, after erection of Bailadila 11 C Project jointly by the official team of NMDC Limited and the equipment suppliers are listed.

16.2 Performance guarantee (PG) test

Salient observations of performance guarantee (PG) test of Bailadila 11 C deposit

1. PG test was conducted between 23rd February '89 and 28th February '89 by an inspection team duly constituted by the Company and the representatives of the Project consultants and

equipment suppliers to ensure achieving specified performance targets as per DPR, Supplier's bids, and performance guarantees.

2. Designed to produce 5 MTPY in Phase I (later on expansion to 7 MTPY in Phase II): This plant could accomplish 5 MTPY design capacity (250 Days/year x 10 hours/day x 2000 TPH).
3. Used three out of four parallel lines at a time (keeping one line as standby to take care of breakdowns, schedule maintenance, and spare capacity to enhance/achieve 7 MTPY capacity in Phase II (provision is made for expansion as and when required with additional equipment if required).
4. Tested to ensure that the total plant as well as individual equipment met the specified duty requirements.
5. Tested and found producing the following specified quality products on a given day (order of magnitude only):

Produ-ct	% Yield	% Fe	% SiO_2	% Al_2O_3	% LOI	% P
Lumps	52.8	64.15	0.83	2.75	4.4	0.022
Fines	34.9	62.16	1.41	4.12	5.31	0.033
Slimes	12.3	59.74	2.95	5.81	5.05	0.065
Feed	100	62.34	1.06	3.82	4.8	0.031

6. Performance of all the individual equipment was evaluated from the amperages drawn on load, abnormal sounds, heating, and so on and found them to be within the norms/acceptable limits.

16.3 Completion Report

Deviations between DPR and Project's Completion Report: Among others, the report should include the following, including deviations, if any, between DPR and final Completion Report:

	As per the Preliminary/ TEFR/DPR reports	**Actual facts/findings after commissioning the plant(CR)**
Ore reserves, grade, ore to waste ratio		
Environmental parameters		
Total capital expenditure in Rupees (Crores)		
Duration of plant implementation (months)		
Total installed power in KVa		
Power consumption in KWH/tonne concentrate		
Total water availability in M^3/hr		
Net water requirement in M^3/tonne concentrate		
Operating cost in Rs./tonne concentrate		
Quality and quantity of tailings		
Tailing dam area and height		
Quantity and quality (physical, chemical, and metallurgical) of finished concentrate		
Marketability/saleability of concentrate		
Causes for unforeseen delays		
Deviations in operating conditions etc.		

The Completion Report would serve as a guide and form the basis for expansion or setting up other similar new plants in the near future with more accurate estimation.

Chapter 17

Estimation of Water Requirement

17.1 Water requirement

Step 1: For any ore processing plant the material balance flowsheet provides key information such as product weight percentage, feed TPH, assay percentage, recovery percentage, solids percentage in pulp, water TPH, Pulp TPH, pulp flow volume (M^3/hr), pulp specific gravity etc. From the material balance flowsheet the following parameters can be calculated.

Total water requirement for ore processing, amount of water recoverable from thickeners (overflow), filters (filtrate), and tailing dams (supernatant water after sedimentation of the solids/tailings) for recirculation.

The net water requirement is the difference between total process water requirement and available quantity for recirculation. This shortfall can be compensated by using water from external sources like bore wells, rivers, streams, rain water catchment ponds, and harvesting pits.

Step 2: For example, the total water requirement for a 1.2 MTPY processing plant working for 6300 hours in a year (315 days/year x 20 hr/day) with an yield of 40% is assumed to be 5 M^3/tonne ROM = 12.5 M^3/tonne concentrate (5 x 100/40).

Assuming the extent of water likely to be recovered from the two thickeners of concentrate and tailings and filter is 90% of total water input; then the net water requirement = 10% of total water requirement (100-90 = 10%) = 0.5 M^3/tonne ROM (5 x 10/100 = 0.5) = 1.25 M^3/tonne concentrate.

Step 3: Total water requirement (for production of 1.2 MTPY)
= 12.5 M^3/tonne concentrate × 1.2 MTPY concentrate
= 15 million M^3/year
= 15 million M^3 per year/6300 hrs. per year
= 2381 M^3/hr
= 2381 × 4.4
= 10,476 gpm
= 15.09 million GPD (10,476 gpm × 60 min/hr × 24 hr/day
= 15,085,440 GPD)

Since 90% water can be recovered for recycling, net water requirement
= 1.25 M^3/tonne concentrate × 1.2 MTPY concentrate
= 1.5 million M^3/year
= 1.5 million M^3/year/6300 hr/year
= 238 M^3 /hr
= 238 × 4.4
= 1048 gpm
= 1.509 million GPD

Hence, the net water requirement of the processing plant alone can be estimated as 1.509 million gallons per day (or 238 M^3/hr).

17.2 Water conservation

As water is a scarce resource, efforts should be made to recover more than 90% total water for recirculation through use of dewatering devices such as thickeners (both for concentrate and tailings), filters, tailing dam, and check dams. The rest of the water is required to be obtained from external sources (rivers, streams, canals, ponds, bore wells etc.). Water conservation measures include having rain water harvesting pits, collection/storage of storm water in tanks/bunds, allowing bore wells to get recharged in turns for continuous water supply, minimizing water evaporation losses, reducing ground water leakages, and so on.

Information source based on consultancy assignment with MSPL Limited gratefully acknowledged.

Chapter 18

Estimation of Power Requirement

Estimating of power requirement: An example is given here for estimation of power requirement. In the absence of power ratings of different equipment used in a process plant, this can be used as a rough guide for estimation of power requirements.

Step 1: Determination of Bond's work index: Bonds work index W_i has to be experimentally determined in the laboratory for any given ore.

Let W_i be assumed as 16 Kwh/tonne .

Step 2: Required grind: It is necessary to undertake ore testing in a laboratory/pilot plant to determine the finest size the ore needs to be ground for achieving separation of the desired constituent from the associated gangue. In other words, this is the extent a particular sized feed sample (size through which 80% feed passes, say F_{80}) has to be ground to a minimum product size (size through which 80% product passes say, P_{80}). This is the required grind for achieving the desired mineral separation, and produce concentrate with the required grade, yield, and recovery.

Step 3: Grinding energy: Calculating grinding energy "W" for reducing a feed sample of a particular size (F_{80}) to the desired product size (P_{80}) can be done using Bond's work index as shown in the following formula:

$$W = 10 \times W_i \times (1/\sqrt{P_{80}} - 1/\sqrt{F_{80}})$$

where

W is the grinding energy required for size reduction of the feed (F_{80}) to required product size (P_{80})

W_i is the Bond's work index of the ore in Kwh/tonne

P_{80} is size in microns through which 80% product passes

F_{80} is size in microns through which 80% feed passes.

For example, the grinding energy required for grinding 80% -6mm feed (-6000 microns) to a ground product of 80% -150 mesh (100 microns) in order to produce a concentrate with an yield of 40% is calculated to be 13.93 Kwh/tonne ROM as shown in the following formula based on the assumed Bond work index of 16 Kwh/tonne ROM:

Grinding energy requirement

$= 10 \times \text{Work index} \times (1/\sqrt{P_{80}} - 1/\sqrt{F_{80}})$

$= 10 \times 16\ (1/\sqrt{100} - 1/\sqrt{6000})$

$= 160\ (1/10 - 1/77.46)$

$= 160 \times 67.46/(10 \times 77.46)$

= 13.93 Kwh/tonne ROM

or 34.825 Kwh/tonne concentrate

$(13.93 \times 100/40)$.

Step 4: Maximum demand. Preliminary estimation of connected load: The correct basis to know the connected load is from the actual bids, indicating equipment power requirements, received from a prospective supplier of equipment for achieving the desired throughput rate of concentrate production. For example, the sum total power ratings of all the equipment as indicated by the equipment supplier is 20 MW. This is the connected load/maximum demand (MD) for producing 1.2 MTPY concentrate per year through processing 3 million tonnes ROM/year with a concentrate yield of 40%. This estimation, arrived from the equipment supplier's bid, would be more realistic as they would take into account safety factors such as motors' efficiency, plant's internal lighting, idle running time due to lack of feed, and so on.

In the absence of the bids from equipment supplier, the following method can be used for a rough estimation of maximum power demand.

As much as 3 MTPY ROM has to be processed in 6300 operating hours in a year at an average feed rate of 548 TPH (3 x 1.15/6300) to produce 1.2 MTPY concentrate of 63% Fe grade. Hence, the grinding power requirement per hour would be 7,634 KW or 7.634 MW as calculated from the Bond's work index (548 TPH x 13.93 kWh/tonne).

METCHEM, former consultants of KIOCL, Kudremukh, using their expertise in providing consultancy for several Indian mineral processing industries, have devised an empirical formula to calculate total energy requirements of a processing plant as well as processing costs. As per this formula, total power requirement of the plant would be twice that of energy required for grinding and, thus, the overall power requirement per hour would be 15,268 kWh (=7634 × 2).

Assuming the power factor to be 0.9, the connected load is estimated to be not less than 16.97 MW (7634 × 2/0.9 = 16,964 Kwh or 16.97 MW). Hence, the maximum power demand of a beneficiation plant designed to produce 1.2 MTPY concentrate of 63% Fe is estimated to be

16.97 MW (548 TPH × 13.93 Kwh × 2/0.9 PF
= 16,964 KW
= 16.97 MW).

Step 5: Energy bill. The annual energy consumption works out to be 92.87 million Kwh for a 1.2 MTPY concentrate production ((13.93 KWh/0.4 yield) × 2 × 1.2 MTPY concentrate)/0.9 PF).

Step 6: Operating cost. Estimation of the order of magnitude of operating costs from the energy consumption can be made using METCHEM's empirical formula.

Assuming the power tariff to be Rs. 6/Kwh, the total annual energy bill for producing 1.2 MTPY concentrate in 6300 hours in a year would be Rs.557.22 million

(92.87 million Kwh × Rs.6/Kwh = Rs.557.22 million).

Based on their wide consultancy experience gained in India, METCHEM, had estimated that the energy bill constitutes 60% of the overall operating cost of an ore processing plant. Based on this thumb rule, annual operating costs of 1.2 MTPY concentrate plant, excluding owning and depreciation costs of the processing plant, would be as follows:

Rs. 928.7 Million (557.22 × 100/60 = Rs.928.7 million).

Hence, the operating cost to produce 1 tonne concentrate from a beneficiation plant having a capacity to produce 1.2 MTPY works out to Rs.773.92 (= Rs.928.7 million/1.2 MTPY) .

Information source based on consultancy assignment with MSPL Limited gratefully acknowledged.

Chapter 19
Equipment and Capital Costs

19.1 Basis of estimation of equipment costs

The equipment cost of a beneficiation plant having the required processing capacity with some built in scope for expansion at 15% has to be worked out based on realistic estimates. Hence, it should be based on the latest bids received for supply of equipment meeting the specified duty requirements. These bids are to be scrutinized to ensure that the equipment offered is new (not reconditioned), reliable, produced by a reputed company, and comes with warranties, process guarantees, assured availability of spare parts and that it enjoys good reputation among other users in the industry. Equipment prices are to be negotiated and purchase orders are to be issued including extra charges payable towards packing and forwarding, sales tax/excise/VAT, service taxes to be paid to State/Central Governments, ocean/railway freight, customs clearance, inland transportation to the site. In addition, the orders should include mutually agreed commercial terms such as equipment warranty/performance guarantees, deliver schedules, inspection clauses and so on. For illustration purposes, the basic cost of plant and machinery for a 300-TPH process plant works out to Rs.50 Crore as shown in Table 19.1.

19.2 Capital costs

Capital costs for the proposed 300 TPH plant has to be worked out carefully to let the promoters/mine owners know the total capital costs/investment amount required for project execution.

19.2.1 Land and site development

In case the land has to be acquired, cost of the land at the prevailing market rates as well as construction of buildings by reputed builders have to be taken into consideration.

19.2.2 Engineering, supervision, and project management

An estimated budget provision (or else, a percentage of the capital cost) has to be made for charges to be expended for design and detailed engineering, supervision, project management, domestic consultancy along with special reports such as environmental impact reports, Government clearances, and so on.

19.2.3 Miscellaneous fixed assets and technology transfer

A realistic budget provision (or else, a percentage of the capital cost) is to be set aside for meeting the costs of process knowhow/technology transfer, royalty, spares, installation, expert consultancy provided by overseas consultants/ equipment erection and commissioning, as well as cost of initial assets, taxes, insurance, and so on.

19.2.4 Initial expenses

Budget provision has to be made for preoperative expenses, including initial expenses, rent, rates, taxes, start up expenses, insurance during construction, and so on.

19.2.5 Debt equity ratio

The debt equity ratio is assumed to be 70:30. Promoters invest equity funds to the tune of 30% of total investment using internal resources and the remaining 70% is borrowed from banks, lending institutions, or internally from its employees, provided the project is profitable and repayment of debt is assured without default and within in a reasonable period.

19.2.6 Contingency expenses

An ad hoc provision is made for the contingency expenses.

19.2.7 Margin money

Margin money amounting to 2% of anticipated annual sales or an ad hoc provision is to be set aside towards meeting working capital requirements. Interest on working capital is to be calculated at the rate of 13% per annum.

19.2.8 Estimation of capital cost

Table 19.2 provides details on capital costs working out to be Rs.75.6 crore based on some assumed figures and percentages or ad hoc provisions that, of course, should not be viewed as a definitive basis or norm applicable to any other project.

Table 19.1: Equipment cost for a 300-TPH iron ore processing plant

Sl. No.	Equipment name	TPH	Size/ source	Quantity required	KW	Total cost (in Rs. lakh)
1	Ore bin (local fabrication)	600 tonnes	300 M^3	1	-	100
2	Grizzly feeder	300	HGF 1248	1	5 × 1 = 5	10 × 1 = 10
3	Double-deck vibrating screen	300	6' W × 16' L	2	10 × 2 = 20	22 × 2 = 44
4	Pressure rolls/rod mills	300	McNally Bharat	2	50 × 2 = 100	200 × 2 = 400
5	Low-intensity wet magnetic separator (LIWMS)	350; 30% solids, 1000 Gauss	600 mm × 2400 mm L, permanent magnetic type (rare earth; Eriez/ Boxmag)	2	10 × 2 = 20	21 × 2 = 42

6	Water pump	200 m^3/hr; head: 20 M	McNally Bharat	4 + 1 stand-by	40 × 4 = 160	30 × 5 = 150
7	Wet high-intensity magnetic separator (WHIMS)	75; 20,000 max. gauss; 2.5 mm gap; 35% solids; 12000 opera-tional gauss; 8R grooved plate	Humboldt Wedag,	4	100 × 4 = 400	375 × 4 = 1500 (to be imported)
8	Slurry pumps	Rubber/ ceramic-lined 200 m^3/hr corrosi-ve solids	McNally Bharat/ Warman	4 + 1 stand-by	50 × 4 = 200	53 × 5 = 265
9	Concentra-te thickener	180 tonnes solids; 35% feed; Under-flow dis-charge: >.50% solids; OF<80 ppm solids	Hindustan Dorr Oliver/ EIMCO KCP	1	10 × 1 = 10	105 × 1 = 105
10	Filter	180 TPH; filter cake: <9.5%	Hindustan Dorr Oliver/ EIMCO KCP	1	105 × 1 = 105	300 × 1 = 300

		moisture; Filterate: <80 ppm solids				
11	0.9m wide belt conveyors (speed: 1 m/second; 10 no.; length 10 metres each;Total length: (10* 10m= 100m)	400 TPH	McNally Bharat	10*50 mL × 0.9 mW	5 × 10 = 50	100×10 = 1000
12	Structural fabrication of chutes, launders, storage tanks, cranes, workshop equipment, and other miscellaneous equipment				200	1084
	Total				1270	3500+1500= 5000

Hence, the estimated cost of plant and machinery for a 300 TPH process plant is Rs.5000 lakh. (Rs.50 Crore) and the connected lode is 1270KW. Caution: The equipment costs provided are approximate and are not realistic.

Table 19.2: Capital costs

Sl. no.	Item	Indigenous source (Rs. in Lakh)	Foreign supplier (Rs. in Lakh)	Total cost (Rs. in Lakh)
1	Land and site development (10 acres at Rs.1 Lakh/acre)	10		10
2	Buildings (assumed 50000 sq ft. x Rs. 2000/sq ft.)	1000		1000
3	Plant and machinery	3500	1500	5000
4	Engineering, project management, and so on at 2% of the cost to be incurred towards building, plant, and machinery	90	30	120
5	Miscellaneous fixed assets	50		50
6	Preliminary expenses at 8% of plant and machinery costs	280	120	400
7	Preoperative expenses at 2% of plant and machinery costs	70	30	100
8	Provision for contingencies at 5% of the cost of indigenous equipment and 10% of the cost of imported equipment	175	150	325
9	Total (1 to 8)	5175	1830	7005
10	Interest during the construction (IDC) period on 50% of total investment cost at the rate of 13% per annum	455.325		455.325
	Total including IDC (Total of items 9 and 10)	7460.325		7460.325
10	Margin money for working capital (ad hoc provision)	100		100
	Grand total	7560.325		7560.325

Total capital investment = Rs.7560.325 Lakh (or Rs.75.6 Crore). The costs provided here are approximate and not realistic.

Information source based on consultancy assignment with MSPL Limited gratefully acknowledged.

Chapter 20

Estimation of Production Cost

20.1 Major steps involved in estimation of production cost

It is necessary to know the information regarding the following parameters before estimating total production cost:

- Available ore reserves of the desired mineral and average grade of the ore.
- Ore to waste ratio.
- Scale of ore production based on market demand, and available resources.

Since setting up a mine along with a beneficiation plant with required infrastructure is quite expensive, the annual throughput rate of production of the mine should preferably be less than 10% of the estimated reserves; that is, mine life should preferably be more the 10 years.

Based on the ore mineral and related parameters such as ore value, associated gangue, and liberation size, a flowsheet illustrating a simple, techno-economically viable process flowsheet should be prepared and used as the basis to arrive at the following test results/data.

Grade of the concentrate yield (i.e., concentrate weight percentage) obtainable under specified operating conditions by the recommended process flowsheet.

Requirement of water, power, and chemicals, including flotation reagents.

To assure the equipment buyers achieve target quality and quantity of saleable products, the prospective equipment supplier/bidder may be advised and persuaded to test the ore sample themselves and give performance guarantees prior to submitting the bid. Otherwise, the manufacturer may supply unreliable equipment even at a cheaper price and try to evade giving performance guarantee, claiming that they are not responsible for reproducing the results obtained by other testing laboratories with regard to grade and yield of the concentrate.

Based on the bids received from equipment suppliers and consultants for design, fabrication, and erection of plant and machinery, the actual equipment cost and capital cost are known.

The market value of the saleable product, tariff rates for water and power consumed, cost per tonne of tailing disposal, and so on have to be ascertained.

Production cost and profitability of the project is worked out on these inputs.

Based on the techno-economics, profit margin, and market demand, decision for going ahead with investing in the project will have to be taken by the Board of Directors of the Company.

20.2 Basis of estimating production cost

Production cost is calculated on the basis of the assumption that plant will produce iron ore concrete at 1.2 MTPY.

Cost of 1 tonne low grade iron ore: Rs.300/tonne

Mine works for 18 hours/day for 300 days/year (5400 hrs/year)

Concentrator works for 20 hrs/day for 315 days/year (6300 hrs/year)

Equipment cost (to be worked out based on actual bids received): Rs.200 Crores

Capital investment (to be worked out based on actuals): Rs.400 Crores

Results of ore testing: Grinding ore to the requisite liberation size of -150 mesh size followed by magnetic separation is expected to produce 63% Fe grade concentrate with an yield of +40%.

Process route includes wet grinding working in closed circuit with cyclones, magnetic separation, dewatering of concentrate and tailings, and filtration of final concentrate to produce filter cake having 9.5% moisture for sale as pellet feed in domestic market.

Connected load (MW) is calculated based on grinding energy requirements, which in this case is assumed to be = 20 MW.

Total power requirement: 30 KWH/tonne ROM; power tariff: Rs.6/Kwh.

Net water requirement: 0.5 M^3/tonne ROM; Water tariff = Rs.10/M^3

Market price of 63% Fe grade iron ore concentrate ex-works: Rs.5000/tonne.

Royalty cost: Rs.500/tonne for <63% Fe iron ore concentrate (10% sale value of Rs.5000/tonne).

Selling expenses at 5% of sale price of iron ore concentrate: Rs.250/tonne concentrate.

Depreciation at 10% on straight-line basis.

Bank interest on long-term loan and working capital at 13% per annum.

Exchange rate 1 US$: Rs.62.

Logistics cost for movement of ore from mine to Process plant = Rs.150/tonne.

Project implementation period: 30 months from the date of investment decision and release of funds.

The tailing dam is proposed to be located close to the process plant, and the cost of slurry transportation from process plant to tailing dam is assumed to be Rs.10/tonne.

20.3 Estimation of the production cost and profitability

Based on these assumptions, the detailed production cost, including royalty per tonne of concentrate ex process plant, has been worked out to be Rs.3972 (US $ 64/tonne concentrate) with a profit of 32% as shown in Table 20.1.

Caution: This figure is not meant to be and should not be taken as the actual production cost. This is just to show an example of the methodology of calculating the production cost based on operating parameters and realistic input costs prior to commissioning the beneficiation plant.

Courtesy: Personal expertise gained from Consultancy assignment with MSPL Limited, Hospet, is hereby acknowledged.

Table 20.1: Estimation of production cost and profitability
(for production of 1.2 MTPY iron ore concentrate)
(all costs in Rs. Million for columns 4, and 7, unless otherwise specified, and except for the last column 9 in US $/tonne concentrate)

Sl. No. (1)	Item (2)	Quantity (3)	Cost Rs. Million (4)	Units required (tonne ROM) (5)	Cost/ tonne ROM (6)	Total annual cost Rs. Million (7)	Cost Rs. / tonne concentrate (8)	Cost US $ / tonne Concentrate (9)
1	Ore reserves (million tonnes)	40						
	Proposed mine life in years	+13						
2. Ore requirement	Required quantity of ROM in MTPY	3						
3	Equipment cost		2000					
4	Capital cost		4000					
5	70% loan capital		2800					
Equity capital	Equity capital at 30% total investment		1200					
6	Percentage of interest	13						
7	Depreciation in years	10						
8	Exchange rate 1 US$ in Rs.	62						

10	Concentrate grade (% Fe)	63						
	Concentrate yield (%)	40						
	Iron ore concentrate (MTPY)	1.2						
	Selling price of 63% Fe grade concentrate (Rs./tonne)	5000						
	Concentrator operating hours/year	6300						
	Purchase cost of Low grade iron ore				300	900	750	12.1
2	Consum-ables (Rs./tonne ROM; flocculant, grinding balls, liner plates, filter cloth, lubricants, etc.)				250	750	625	10.08
3	Utilities							
	Energy cost (Rs./kwh)	6		30 Kwh/ tonne ROM	180	540	450	7.26
	Net water cost (Rs./M^3)	10		0.5 M^3/ tonne ROM	5	15	12.5	0.2

4	Cost of wages (Rs. Million): 30/shift × (3 shifts + 1 Relief shift = 4 shifts) × 15000/person × 12 months/year				7.2	21.6	18	0.29
5	Repairs and maintenance at 10% of capital cost	10%	4000		133.3	400	333.33	5.38
6	Overheads at 10% of sum of above items	10%			87.55	262.66	218.88	3.53
7	Logistic cost of ore movement from mine to process plant at Rs.150/tonne ROM	Rs.150/ tonne ROM			150	450	375	6.05
	Tailings disposal through slurry transportation for 5-km distance from process plant to tailing dam	Rs.10/ tonne tailing			6.00	18.0	15	0.24

8	**Total operating cost A (Rs./tonne)**				**1119.1**	**3357.3**	**2797.7**	**45.1**
9	Average interest spread over 10 years on loan capital repayable in 5 years	13% on 2800 Million loan capital	109.2		36.4	109.2	91	1.5
10	Depreciation at 10% on total investment of Rs.4000 million	10	4000		133.3	400	333.3	5.4
11	**Total owning cost B**				**169.73**	**509.2**	**424.3**	**6.84**
12	**Total production cost = operating cost (A) + owning cost (B)**				**1288. 8**	**3866.5**	**3222.0**	**52**
13	Selling price of iron ore concentrate						5000	80.7
	Selling expenses at 5% of the sales value of concentrate	5% sales price			100	300	250	4.0
	Royalty at 10% sales value of concentrate	10% sales price			200	600	500	8.1

	Total production cost, including selling expenses and royalty				**1588.8**	**4766.5**	**3972.0**	**64.1**
14	Profit/loss margin for selling in domestic market						1028	16.6
	Percentage of profit over production cost						**32**	**32**

Information source based on consultancy assignment with MSPL Limited gratefully acknowledged.

Chapter 21
Environmental Issues

21.1 Nature/type of pollution issues

In general, pollution issues include:

- Pollution of water by rain (benches of ore deposit, waste dumps, tailings, and so on getting washed out by rain).
- Pollution of air due to effluent gases such as carbon dioxide, carbon monoxide, and sulphur dioxide (caused by roasters, smelters, and reduction kilns).
- Dust pollution of air with ultra fine dust particles due to dry ore processing (e.g., barytes, mica, silica mines, etc.).
- Sound pollution due to operation of ore processing equipment such as crushers, grinding equipment, sinter strands, pellet heat hardening furnaces, and so on.

21.2 Tailing disposal/conveyance to dam

- It is desirable to locate the tailing dam away from the processing plant/human habitat for ensuring clean environment and potable water availability for inhabitants living in the vicinity. Tailings slurry containing around 10 to 20% solids can be dewatered in a tailing thickener to more than 50% solids before its rejection into settling ponds/tailing dam for water recovery. The tailings can be rejected into the dam through any one of the following alternate methods:
- If water availability is not a constraint, tailings can be discarded by gravity through an open channel, or drain pipes, provided it does not intersect any road (can preferably be designed to flow along the length of the road).
- The preferred method is slurry transportation through a pipe line at an inclination of at least 1 degree. Furthermore, it must be ensured

that slurry pipes are laid above the ground surface to facilitate easy inspection/maintenance.

- Back filling the excavated mine pits with tailings can be considered, provided the ore is totally removed from the mine benches/pits.
- Tailings can be dewatered in a thickener, followed by rejection of the thickened solids into different storage tanks (keeping a spare tank/pond for collection is essential, if other tanks are filled to the brim and need to be evacuated periodically).
- Tailings can be pumped away to a dam located far away in an isolated zone, if space and suitable valley like topography is available.
- Tailings can be subjected to cyclone treatment, and the cyclone overflow can flow by gravity into a valley/dam . After further treatment, the cyclone underflow can be used as a saleable product as and when there is market demand.

21.3 Rough estimation of space requirements of tailing dam

Assumptions: Mine excavation: 5 MTPY ROM

Life of the mine: 10 years

Weight percentage of tailings (cyclone overflow): 10% of ROM = 0.5 MTPY

Bulk density = 2.5 tonnes/M^3

Safe height of tailings that can be stored in the dam: 10 metres

Quantity of tailings to be rejected/required storage capacity in the dam for the mine life = 10% of 5 MTPY × 10 years = 0.5 MTPY × 10 years = 5 million tonnes solids

Required volume of the dam = 5 million tonnes/2.5 tonne per M^3 = 2 million M^3

Total area of the tailing dam = 2 million M^3/10 m (height) = 2×10^5 sq. metres. = $2 \times 10^5/10^4$ = 20 hectares

Giving a safety margin of 20%, total minimum area required for tailing dam = 20 × 1.2 = 24 hectares

21.4 Commercial utilization of tailings to improve profitability

Promising avenues to reduce tailings/solid pollutants are as follows:

1. Quantity of tailings can be reduced by recovering additional values from the existing dumps (like possibility of gold recovery from the existing Kolar Gold Fields or Hutti Gold Mines or selling the tailings as and when there is increased demand. In fact most iron ore mines cleared their stocks of low grade dump fines and sold them in the export market.
2. Trying to sweeten the low grade dump fines with fresh high grade concentrate and trying to use it for pellet making/iron making (like Donimalai Iron Mine is wisely planning to do for value addition).
3. Commercial use of tailings for making bricks, pavement blocks, road tiles, and so on, which will not only control pollution but also generate additional employment, a socioeconomic responsibility for the industries.
4. In China, the high silica iron ore tailings are reportedly used for paving roads.
5. Recovery of molybdenum, nickel, gold, and silver from the tailings if available, by using latest processing techniques (like the one adopted for extracting copper concentrate, in Ghatsila mines).

21.5 Pollution/environmental standards

Ministry of Environment and Forestry issues from time to time certain standards to be adhered to by all industries strictly, and any industry that fails to adhere to the standards may be forcibly closed down.

Suspended solids in water should be less than 80 ppm.

The standards for drinking water are as follows:

Requirements/specifications for drinking water/(desirable limits) as per IS 10500-19, edition 2.2 (200320-09)

Chemical constituents in water	Desirable limits
Total hardness (as $CaCO_3$) mg/lit. max	300
Total dissolved solids mg/lit. max	500
Iron (as Fe) mg/lit. max	0.3
Chloride (as Cl) mg/lit. max	250
Nitrate (as NO_3) mg/lit. max	45
Fluoride (as F) mg/lit. max	1.0
Mercury (as Hg) mg/lit. max	0.001
Arsenic (as As) mg/lit. max	0.01
Lead (as Pb) mg/lit. max	0.05
Pesticides mg/lit. max	Absent

Sound in the factory/mine/ or processing plant should not exceed a certain decibel level.

Effluent gases from the mine should be free from harmful gases (such as CO, CO_2, SO_2, and suspended dust particles in the air).

The overflow water from the tailing dam should be pre-treated to ensure concentration of objectionable chemicals (like Na CN, acids) below the set safe limits for humans and animals.

21.6 Issues in 'illegal mining'

Production and transportation of iron ore out of the mines located in the states of Karnataka, Goa, Odisha were totally stopped/suspended in the financial year 2012-2013, except the mines of NMDC Limited (a public sector undertaking), by the Supreme Court/Government of India. The reasons were that these mines were violating several statutory rules, such as mining much in excess of their sanctioned mine plan and exporting ore without prior sanction by all means available and making huge profits. The other factors include environmental pollution, encroachment of forest land, conflict with local population/tribals and farmers over their land rights, and severely damaging the roads by plying overloaded trucks carrying iron ore fines with spillages. Furthermore, high export demand, even for the low grade iron ore fines (say 55% Fe), had prompted some miners to export in violation of all statutory regulations. Whether these deviations can be viewed as illegal or unscientific mining is a point of debate. This has crippled the export trade and Indian economy, leading to widespread unemployment, slowdown of iron ore industries, such as pellet plants, sponge iron plants, mini steel plants, owing to the lack of basic raw material. This is creating a peculiar situation where one is forced to import ore even though the same is available locally but could not be mined due to Court orders. For example, steel plants are forced to import iron ore and coking coal at huge costs or acquire mining properties abroad, though the resource thus imported is in some cases available in abundance locally and in the importers' own deposits/captive mines.

21.7 Two examples of projects totally stopped due to concerns of environment

1. The Kudremukh Iron Ore Company of Chickmagalur, Karnataka, has more than 400 million tonnes magnetite/hematite balance ore reserves. Because of committing violations it had to be shut down in 2005.
2. Bababudan, another promising iron ore deposit, has more than 600 million tonnes of proved iron ore reserves but it could not get Government clearances for mining due its location/proximity to high value coffee estates reserve forest area with bamboo trees, wildlife game sanctuary, and so on. Plantation owners opposed mining on the grounds that coffee revenue is a perpetual high value agricultural income against depleting wealth of iron ore reserves if mining is allowed from the location.

Chapter 22

Site Selection for Setting up an Ore-Processing Plant

22.1 Issues to be considered for setting up an ore processing plant

1. *Minimal distance between the mine and process plant:* It should preferably be close to the mine/source of raw materials as well as other accessory ores like coal (for generating thermal power), fluxes (limestone, dolomite, Dunite), binders (bentonite), besides storage space for stocking at least a week's ore requirement for blending and transporting the ore to the crushing plant. In case the mine is located within the forest area, it would be difficult to get the necessary clearances for locating the process plant within the mine/lease area. Available options for movement of ore from mine to processing plant are as follows:
   - Truck transportation
   - Railways, provided there is the railway sidings both at the mine and at the process plant
   - Aerial ropeway, provided crushing is done at the mine
   - Slurry transportation, an option suitable only if both crushing and grinding are done at the mine

2. Adequate land has to be procured for immediate requirement as well as future expansion plans of the plant and to accommodate all the following essential facilities:
   - Weigh bridges (one for incoming raw material at the entrance, and the second placed ahead of the exit gate for the finished concentrate/saleable products, and disposable waste solids (like tailings)
   - Storage and blending yard for incoming raw ore
   - Crushing, screening, and processing sections
   - Dewatering and drying sections

- Concentrate/finished product storage shed
- Water-receiving tank and water recirculation and clarified/clean water tank
- Space for internal truck movement
- Maintenance of heavy earth moving equipment and workshop
- Railway siding (if feasible/required/and justified)
- Thermal power station/standby generator
- Dust collection/disposal equipment
- Post office/banks/recreation clubs/playground, schools, garden and so on
- Other essential amenities (security, canteen, dispensary/first aid room, residence for emergency staff like pump operator, electrician, welder, ambulance driver, etc.)

3. Power: Availability of power from a nearby HT supply: Capital costs go up with the increased distance for laying the HT supply lines.
4. Water supply: Source for the required quantity of water should be available either within the process plant (like tapping water from a nearby river/dam/bore wells/purchased water in tankers, etc.) or brought in from the nearest available river. Running costs for pumping water goes up with an increase in the pipe line distance.
5. Tailing dam: It is necessary to find suitable space for locating a dam, like a valley surrounded by hills on three sides, for storage of tailings likely to be generated over the life of the mine and recovery of water for its recirculation to cut down net water requirement/save precious water besides pollution control.
6. Availability of market for the produced concentrate nearby: Though the concentrate is sold at ex- works cost basis, shorter distance between the process plant and the user industry (smelter/pellet, sponge iron, mini steel, and railway siding) reduces transportation costs and results with availability of concentrate at lesser landed price to the actual buyer or user.

7. Closeness to the railway station: For getting the raw ore and dispatch of the final concentrate to the user industry, closeness to the railway station reduces the transport costs.
8. Proximity to towns/habitat: It is desirable for the personnel working in the process plant to stay close to the plant, as they are expected to work in shifts around the clock and need to attend to any unforeseen plant breakdowns at short notice. If the process plant is close to the habitat/town, separate residential accommodation may not be needed, which may help company save on capital costs.
9. Closeness to the highway/bus station/railway station and airport: This is to create better connectivity/approach for the employees so they can travel easily between their hometown and work area.
10. Hospital with all the essential/emergency facilities and blood bank: This is required to provide better healthcare to the employees, their families, and the local people.

Appendix 1 : Formulae Commonly Used in Mineral Processing

1. **Mass, Density, specific gravity (sp.gr), flow rates**

   Density: Mass / volume = grams/c.c. = kg/litre

   Specific gravity: Ore density / water density

   Bulk density of the ore: Ore mass / bulk volume = kg/cubic meter

   Pulp density = Pulp mass / pulp volume = Mass of 1 litre pulp

   Perentage of solids in the pulp: Mass of the solids in I litre pulp x 100 / Mass of 1 litre pulp

   TPHS = Tonnes solids/hour = Rate of flow of solids in the system

   TPHW = Tonnes water/hour = Rate of flow of water in the system

   TPHP = Tonnes pulp/hour = Rate of flow of pulp = (TPHS + TPHW)/hour

   Flow rate of pulp in M^3/Hr = ((TPHS / sp.gr of solids) + (TPHW / sp.gr. of liquid))

   Sp.gr. of pulp = TPHP / pulp volume in M^3 per hr
   = TPHP/((TPHS/sp.gr of solids) + (TPHW/ sp.gr. of liquid))

2. **Crushing operations:**

   Reduction ratio = d_{80f} / d_{80p}

   where d_{80f} is the screen size through which 80% feed passes and d_{80p} is the screen size through which 80% of the product passes.

   Crusher capacity as per Taggart formula: $T_t = w\,(o + s) \times K_t/2$

   where T_t = Tonnes/hour; w: width of receiving opening in inches; o: open setting in inches; s: closed setting in inches; K_t: constant

   Crusher product throughput rate: $T_R = T\, R_{80}.K.k'.k''$

   where T_R: Crushed product in tonnes/hour; T: feed tonnes/hour; R_{80}: 80% reduction ratio; and k, k' and k" are constants

K : constant (depends on crushability of rock; assume 1 for limestone and 0.8 for granite)

K' : constant (depends on moisture; assume 1 for dry feed)

K" : Constant (depends on feed/crusher operating size; assume 0.75 to 1 for laboratory jaw crusher)

3. **Yield %, recovery from three product assays**

Weight percentage or Yield percentage

= Mass of the concentrate x 100 / mass of the feed

= 100 C/F = 100 × (f-t) / (c- t)

Recovery percentage of valuable mineral

= Cc x 100 / Ff = c(f – t) × 100 / f(c-t)

where C = concentrate mass;

F = feed mass;

'f', 'c', and 't' are the assay values of the desired element pertaining to the feed, concentrate, and tailings, respectively.

4. **Work index, grinding energy, order of magnitude of the cost of ore processing:**

Grinding energy requirement / ton ROM

$= W = 10 \times W_i \times (1/\sqrt{P_{80}} - 1/\sqrt{F_{80}})$

Where W_i is the Bond's work index;

'P_{80}' and 'F_{80}' are the screen sizes in microns through which 80% of product and 80% of the feed passes.

Total energy required for processing / ton ROM as per METCHEM's empirical formulae

= 2 × grinding energy requirement / ton ROM

Total energy required/ton concentrate

= (2 × grinding energy requirement per ton ROM) × 100 / concentrate yield

Empirical formula to know the order of magnitude of the cost of processing / ton concentrate

= 1.67 × power tariff rate × total energy consumption/ton concentrate

= 1.67 × power tariff rate per unit KWH × total energy requirement for processing / ton concentrate.

5. Filter press design calculations:

Assumptions for filter design (for illustration only):

Required output of filter cake = 300 TPH

Plant running hours = 20 hrs/day

Desired moisture content = 9.5%

Cake dryness = 100 - 9.5 = 90.5% = 0.905

Cake density (assumed) = 2.5 tonnes/M^3

Cycle time = 18 minutes

Squeezing factor = 0.9 cycle time

No. of cycles/day = 20 hrs per day × 60 min per hr / 18 min per cycle = 66 cycles/day

Volume of filter press = Total weight in tonnes output per day / (cake density × squeezing factor x dryness of cake × cycles per day)

= (300 TPH × 20 hrs/day) / (2.5 × 0.9 × 0. 905 × 66)

= 6000 / 134.39 = 44 M^3 = 44000 litres/day

Hence, the volume of a filter press working 20 hrs/day requiring to produce, through filtration, 300 TPH cake of 9.5% moisture = 44000 litres/day

6. Slurry pump calculations

Pulp velocity through pipe = Q / 1000A = Q / 785 D^2

Velocity head = $v^2/2g$

Water horse power = BHP = Q × H × G / 75

Pump efficiency = n = W × H × P / BHP

Electric horse power = HP = BHP × 1000 / motor efficiency

where Q = pump discharge in litres/second

D = diameter of the pipe in meters

A = cross-sectional area of pipe in sq. mtr. = $\pi r^2 = \pi D^2/4$

v = pulp velocity in meters/second

Head = H = friction head

g = acceleration due to gravity in m/sec^2

= 9.81 m/sec^2 = 981 cm/sec^2

G = specific gravity of fluid to be pumped

W = specific weight of fluid (for water, 1000 kg/m^3)

= 1000 × G

L= length of pipe in meters

F = Luminous friction coefficient

HP = 746 watts

Pump efficiency = say 85% = 0.85

Motor efficiency = say 80% = 0.8

Break horse power of the pump

= (Q × G × H × D) / (75 × pump efficiency × motor efficiency)

7. Screening capacity

Screening capacity/sq.ft. = Lbs per cft. × stone capacity in tons per hour/sq.ft, through opening /100

Where stone capacity in tons per hour/sq.ft. through the desired opening size is available from equipment manufactures catalogue.

Or screen capacity in

TPH = stone capacity (in TPH/sq.ft.) × screen area (sq.ft.) × bulk density of ore/bulk density of rock (=2.65)

Appendix 2 : Genesis of Rocks and Mineral Deposits

Genesis and classification of rocks:

Rocks are broadly divided into three broad categories:

Igneous rocks: About 95% of earth's crust is composed of igneous rocks and formed due to crystallization of molten magma.

Sedimentary rocks: Most of the igneous rocks are blanketed by sedimentary rocks.

Metamorphic rocks: Complete or partial replacement of the pre existing rocks in the earth's crust occurs when they are subjected to high temperature and pressure.

Igneous rocks: They are formed due to solidification of molten magma coming with great force from earth's interior before it reaches the surface. The solidified rocks are classified differently, based on the zone of solidification, silica percentages, and mineral assemblages based on the composition of molten lava and so on as shown below:

Classification of rocks based on the zone of solidification:

- Intrusive : The rocks are solidified below the earth's surface
- Extrusive : The rocks are formed due to solidification of magma on the earth's surface

The ore deposits are also classified based on their stages of consolidation/formation.

1. Ortho-magmatic stage: On cooling, first, the non volatile constituents of magma form as the rock forming silicates in ortho - magmatic stage.
2. Pegmatitic stage: The liquid part of the remaining magma gets solidified in pegmatitic stage forming pegmatites.
3. Pneumatolytic stage: The gaseous part of magma gets solidified in pneumatolytic stage.

4. Hydrothermal stage: The last part of consolidation occurs along with the hot water/rich solutions in hydrothermal stage.

Classification of igneous rocks based on its silica content:

1. Acidic rocks: > 66% SiO_2
2. Intermediate rocks: 52 - 66% SiO_2
3. Basic rocks: 45 – 52% SiO_2
4. Ultra basic rocks: < 45% SiO_2

Both mineralogy and chemical composition of the freshly formed igneous rock depends on the original composition of the magma and its rate of cooling, besides temperature, pressure, and composition of country rock hosting the solidification.

Formation of minerals based on original composition of the magma:

1. If the original magma is rich in alkalis, silica, alumina, then minerals such as feldspar, quartz, mica, and so on are formed.
2. If the original magma is low in silica but high in alkalis, feldspathoids are formed.
3. If the original magma is rich in lime, magnesia, and iron, then ferro-magnesium minerals are formed.

The igneous rocks are classified based on the association of minerals present in the rock:

Category A: rocks containing quartz and feldspar: granites containing quartz, orthoclase with or without mica, and granodiorite containing quartz, plagioclase, mica with or without hornblende

Category B: rocks without quartz but with feldspar:
Syenite with alkali feldspar, with/or without mica and hornblende
Diorite with plagioclase with or without mica and hornblende

Gabbro, dolerite with plagioclase, and pyroxene with or without olivine
Feldspar and feldspathoids

Category C: rocks containing mostly ferro-magnesium minerals like olivine, pyroxene, or hornblende.

Category D: pyroclastic rocks/fragmented rocks formed due to violent eruption: The magma is blown into pieces along with the host country rocks forming agglomerates (fragmental rocks), tuff, and ash (fine grained material).

Sedimentary rocks: These are formed due to the deposition of material arising from the physical and chemical decay of pre-existing rocks on the earth's surface. These rocks disintegrate over long periods of geological time by the agents of mechanical and chemical weathering, such as frost, rain, ice, water, wind, and chemical decay. These minerals move from its place of origin and get deposited elsewhere at a new / alternate / and more convenient location.

These sedimentary rocks are divided based on their origin:

A. Detrital: Sedimentary rocks are formed due to the deposition of transported particles. They include pebbles, sandy particles, and clay material.
B. Chemical / organic: These rocks are formed due to precipitation of substances from solution or by organic processes. These sedimentary rocks contain carbonaceous, siliceous, ferruginous, aluminous, and phosphoric substances.

Metamorphic rocks: Complete or partial replacement of the pre-existing rocks in the earth's crust occurs when they are subjected to high temperature and pressure, followed by migration of material. Under the new environment, new minerals are formed in place of unstable original minerals. Metamorphic rocks also contain quartz, feldspar, micas, pyroxenes, and hornblende just as in igneous rocks. The new mineral composition of metamorphic rock depends on the composition of original rock and conditions of metamorphism. Typical metamorphic minerals are garnet, andalusite, kyanite, sillimanite, talc, chlorite, epidote, and so on.

Mineral deposits: A mineral deposit contains a valuable element or mineral in sufficient quantity associated with some rejectable gangue worth exploiting profitably. Mineral deposits are classified based on their formation with respect to the surrounding host rock:

- Syngenitic deposits: These are formed at the same time as the associated rock due to segregation in the orthomagmatic/first stage of magma solidification (e.g., chromite in ultrabasic igneous rocks).
- Epigenetic deposits: These deposits are formed later due to subsequent volcanic activity, in the form of veins or lodes, or as interstices in the associated host rock. In some cases, they replace the host rock. Mineral deposits may also undergo metamorphism retaining some characteristics of the host rock.

Types of mineral deposits: Concentration of an element in a deposit is necessary before it can be mined profitably. It occurs in two different environments:

A. Endogenetic igneous deposits: These form within the earth's crust due to solidification of molten magma during the progress of its cooling at different temperatures as shown below:

I. Igneous or magmatic deposits:

1. Orthomagmatic deposits form between 700 and 1500°C resulting with formation of important deposits of native metals (platinum), metallic oxides (magnetite, ilmenite, and chromite), and sulphides (chalcopyrite, pyrrhotite)
2. Pyrometasomatic deposits form between 500 and 800°C resulting with formation of sulphides (chalcopyrite, pyrite, zinc blende) and oxides (such as magnetite, hematite, iron garnets, wollastonite, epidote, etc.).

3. Pegmatitic deposits form at around 575°C resulting with formation of large-sized crystals of feldspar, quartz, mica, gem stones, tourmaline, topaz, and so on.
4. Hydrothermal deposits form at around 50–500°C resulting with formation of large deposits of gold, lead-zinc, copper, mercury, and so on.

II. Metamorpic deposits: Deposits of hematite and magnetite may also form due to metamorphism of impure hydrated iron ores, graphite, asbestos, and so on at around 400°C:

B. Exogenetic sedimentary deposits: These are formed due to physical/surface processes such as deposition of a mineral in a salt lake or sorting by river flow. Sedimentary mineral deposits can be divided into three categories:

1. Weathering residual deposits: They are remnants of the main rock formed upon removal of gangue minerals from the original rock by weathering (e.g., gold, tin, and bauxite).
2. Detrital, alluvial, or placer deposits: They are formed by stream or wave action at low velocities. (e.g. gold, platinum, gems, tin, wolframite, magnetite, zircon, chromite, etc.). Alluvial deposits result due to breaking up of the parent rock, subsequent transportation of valuable minerals, and concentration of such minerals at different locations.
3. Chemical (organic) mineral deposits: Limestone, dolomite, iron ores, copper ores, saline residues, phosphates, coals, and so on are formed due to chemical or organic reaction of the original rock formation.

Supergene enrichment: Ore deposits undergo weathering and decomposition of mineral outcrops (called gossan). Native metals frequently occur in gossan due to removal of lighter gangue minerals. If galena is present at depth, cerussite can be expected to be present in the gossan.

Secondary enrichment of supergene deposits: It can result with formation of silver, copper, and other deposits. The oxy-salts produced in gossan by weathering are carried deeper into the lode located below the earth's crust by descending waters. Here chemical reaction takes place between descending waters carrying oxy - salts and unaltered sulphides located deep below forming enriched metals/ lodes suitable for commercial/ profitable mining.

Appendix 3 : Important characteristics of some economic ore minerals

Mineral	Formula	% element	Hardness	Sp.gr	Floatable property – natural floatability (NF); floatable (PF); difficult to float (DF)	Magnetic property- strongly magnetic: SM; weakly magnetic: WM; sometimes magnetic: STM	Response to HTS separation-'T' thrown from HTS or 'P' pinned down to HTS
Anglesite	$PbSO_4$	Pb:68.3	2.8-3	6.1-6.4			
Apatite	$Ca_4(CaF)(PO_4)_3$	P_2O_5:42.3	4.5-5	3.2			P
Argentite	Ag_2S	Ag:87.1	2-2.5	7.2-7.4			
Arsenopyrite	FeAsS	As:46	5.5-6	5.9-6.3			
Barytes	$BaSO_4$	BaO: 65.7	2.5-3.5	4.3-4.6			P
Bauxite	$Al_2O_3.3H_2O$	Al:34.9	1-3	2.6			
Bentonite	$(CaMg)O.SiO_2(AlFe)_2O_3$	Clay	1	2.1			
Beryl	$Be_3Al_2(SiO_3)_6$	Be:5	7.5-8	2.6-2.8			
Biotite	$(HK)_2(MgFe)_2Al_2(SiO_4)_3$	Mica	2.5-3	2.7-3.1			
Bornite	Cu_5FeS_4	Cu: 63.3	3-3.5	4.9-5.4			

Calcite	$CaCO_3$	CaO:56	3	2.7			P
Cassiterite	SnO_2	Sn:78.8	6-7	6.8-7.1			T
Cerussite	$PbCO_3$	Pb: 77.5	3-3.5	6.5-6.6			
Chalcocite	Cu_2S	Cu:79.8	2.5-3	5.5-5.8			
Chalcopyrite	$CuFeS_2$	Cu:34.6	3.5-4	4.1-4.3			
Chromite	$FeO.Cr_2O_3$	Cr:46.2	5.5	4.3-4.6			T
Cinnabar	HgS	Hg:86.2	2-2.5	8-8.2			
Coal	C				NF		
Cobaltite	CoAsS	Co:35.5	5.5	6-6.3			
Columbite	$(FeMn)(NbTa)_2O_6$	Variable	6	6.3			
Corundum	Al_2O_3	Al:52.9	9	3.9-4.1		WM	P
Covellite	CuS	Cu:66.5	1.5-2	4.6			
Cryolite	Na_3AlF_6	F:54.4	2.5	3.0			
Cuprite	Cu_2O	Cu:88.8	3.5-4	5.9-6.2			
Diamond	C	C:100	10	3.5			T
Dolomite	$CaMg\ (CO_3)_2$	MgO:21.9	3.5-4	2.8-2.9			
Ferberite	$FeWO_4$	W:60.6	5-5.5	7.2-7.5			
Fluorite/Fluorspar	CaF_2	F:48.9	4	3-3.3			T
Franklinite	$(ZnFeMn)O(FeMn)_2O_3$	Zn:14.2	5.5-6.5	5.2		SM	

Galena	PbS	Pb:86.6	3	7.4-7.6			T
Garnet	vary		6.5-7.5	3.2-4.3		WM	P
Gibbsite	$Al(OH)_3$	Al:34.6	2-3.5	2.4			
Gold	Au	Au:100	2.8	15.6-19.3			T
Graphite	C	C:100	1-2	2.2	NF		
Gypsum	$CaSO_4.2H_2O$	CaO:32,6	1.5-2	2.3			P
Halite	NaCl	Na:39.4	2.5	2.1-2.6			
Hausmannite	Mn_3O_4	Mn:72	5.3	4.7			
Hematite	Fe_2O_3	Fe:70	5.5-6.5	4.9-5.3		WM	T
Ilmenite	$FeTiO_3$	Ti:31.6	5-6	4.5-5		SM	T
Kyanite	Al_2O_3 SiO_2		3.23-3.25	6.6-7.5			P
Limestone	$CaCo_3$	Ca:40	3	2.7			
Limonite	$2Fe_2O_3.3H_2O$	Fe:59.9	5-5.5	3.6-4		WM	T
Magnesite	$MgCO_3$	Mg:28.9	4-4.5	3.1			
Magnetite	$FeO.Fe_2O_3$	Fe:72.4	5.5-6.5	5.2		SM	T
Malachite	$CuCO_3.Cu\ (OH)_2$	Cu:57.5	3.5-4	4			
Marble	$CaCO_3$	Ca:40	3	2.7			
Molybdenite	MoS_2	Mo:60	1-1.5	4.7-4.8	NF		
Monazite	(CeLaDy) PO_4ThSiO_4	ThO_2:9	2-2.5	2.8-3		WM	P

Muscovite	$H_2KAl_3(SiO_4)_3$		2-2.5	2.8-3			
Niccolite	NiAs	Ni:44.1	5-5.5	7.3-7.7			
Pentlandite	(Fe Ni)S	Ni:22	3.5-4	4.6-5			
Platinum	Pt	Pt:100	4.5	17		STM	
Psilomelane	$MnO_2,H_2O.$ K_2BaO_2		5-6	3.7-4.7			
Pyrargyrite	$3Ag_2S.Sb_2S_3$	Ag:60	2.5	5.8-5.9			
Pyrite	FeS_2	Fe:46.7	6-6.5	5			T
Pyrolusite	MnO_2	Mn:63.2	1-2.5	4.8		WM	
Pyrrhotite	Fe_5S_6 to $Fe_{16}S_{17}$	variable	3.5-4.6	4.6		WM	
Quartz	SiO_2	Si:46.9	7	2.65-2.66			P
Realgar	AsS	As:70.1	1.5-2	2.6			
Rhodochrosite	$MnCO_3$	MnO:61.7	3.5-4.5	3.5-3.6			
Ruby	Al_2O_3	Al:52.9	9	4			
Rutile	TiO_2	Ti:60	6-6.5	4.2			T
Scheelite	$CaWO_4$	W:63.9	4.5-5	5.9-6.1			P
Siderite	$FeCO_3$	Fe:48.3	3.5-4	3.9		WM	
Silver	Ag	Ag:100	2.8	10.5			
Sillimanite	$Al_2O_3SiO_2$		3.23-3.25	6.6-7.5			P
Sperrylite	$PtAs_2$	Pt:56.6	6.6	10.6			

Sphalerite	Zn S	Zn:67.1	3.5-4	3.9-4.1			T
Spinel	$MgO.Al_2O_3$	MgO:28.2	8	3.5-4.1			P
Stibnite	Sb_2S_3	Sb:71.8	2	4.5-4.6			T
Strontianite	$SrCO_3$	Sr:59.3	3.5-4	3.7			
Sulphur	S	S:100	2.05-2.09	1.5-2.5	NF		
Sylvanite	$(AuAg)Te_2$	Au:24.5	1.5-2	7.9-8.3			
Talc	$H_2Mg_3(SiO_3)_4$	Mg:19.2	1	2.7-2.8	NF		
Tantalite	$FeTa_2O_6$	Ta_2O_5 :65.6	6.3	5.3-7.3			T
Tourmaline	$(NaLiK)_6$ $(MgFeCa)_3$ $(AlCrFe)_2$ B_2SiO_5	No metal	7-7.5	3-3.2			P
Uraninite	UO_3,UO_2	Radium source	5.5	9-9.7			
Wolframite	$(Fe\ Mn)WO_4$	W:51.3	5-5.5	7.2-7.5			T
Zincite	ZnO	Zn:80.3	4-4.5	5.4-5.7			
Zircon	$ZrSiO_4$	ZrO_2:67.2	7.5	4.2-4.7		WM	P

Appendix 4: Indian mineral reserves (Statistical Profiles of Minerals 2010-2011 by IBM)

Ore/mineral	Reserves in million tonnes (as on 1st April 2010)	Life index (years)	Grade percentage	India's annual production MTPY	India's position in the world
Apatite	24.229	Very large	16-20 P_2O_5	0.003845	
Barytes	72.734	16		2.334	III
Bauxite	3479.62	181	40-50% Al_2O_3	12.641	VI
Calcite	20.945	432		0.039370	
Limestone (as on 1st April 2005 data)	175,345	270		237.774	
Chromite	203.346	29	Mostly <52% Cr_2O_3	4.262	III
Copper ore/concentrate	1558.458	225	Ore: 0.88 to 1.% Cu Concentrate: 17-26% Cu	3.615 ore; 0.137 concentrate	
Diamond	31.922 million carats	144		0.019774	
Dolomite	7731	471		5.065	
Fluorite	18.214	Very large		Graded: 0.003150; Concentrate: 0.004394	

Garnet	56.963	14		2.058	
Gold	Ore: 519.816; Metal: 0.000666	190		Ore : 0.000727; Gold: 0.000,002.239	
Graphite	174.85	168		0.114836	
Gypsum	1286.498	196		4.347	
Iron ore	28526	73	60-65% Fe saleable products	207.998	
Hematite	17882		Mostly 60-65% Fe		IV
Magnetite	10644				
Kyanite	103.246	Very large	66% production of <40% Al_2O_3	0.005569	
Lead zinc concentrates	685.595	47	Lead and zinc Ore: 2.03% Pb; 11.54% Zn; Lead concentrate: 57.46 Pb Zinc concentrate: 51.16 Zn	Ore:7.490; Lead concentrate: 0.145 Zinc concentrate: 1.420	
Magnesite	335.172	Very large		0.23	
Manganese ore	429.98	94	25-35% Mn	2.881	VI
Mica	0.532237	337		0.001293	I
Pyrites	1674				

Silica sand	3499	445		3.081	III
Silver	Ore: 466.985; Silver metal: 0.028	105		0.148288	
Sillimanite	66.987	Very large		0.047671	
Sulphur	0.00021			0.237	
Tin concentrate	Ore: 83.726197; metal: 0.102275	Very large		61.355	
Tungsten concentrate	Ore: 87.387; Metal: 0.142				

(Information source: Statistical profiles of Minerals 2010-2011; Indian Bureau of Mines: gratefully acknowledged)

Appendix 5: Commonly Used Conversion Factors

Length	1 metre	= 39.37 inches	= 3.2808 feet = 1.094 yard	= 100 cm	= 1000 mm
	1.6093 km		= 1 mile		
	1853.27 metre = 1.85327km		1 nautical mile = 1.1516 mile		
		1 yard = 3 feet = 36 inches = 0.9144 metre			
		1 inch		= 2.54 cm	= 25.4 mm
			1 feet = 0.3333 yard = 0.3048 metre	= 30.48 cm	1 mm = 1000 microns
			1 feet	= 304.8 mm	10 mesh = 1680 microns
					100 mesh = 150 microns
					150 mesh = 100 microns
					200 mesh = 74 microns
					325 mesh = 44 microns

Area	1 sq. meter		=1.196 sq.yd. =10.76 sq.ft.	= 10^4 sq.cm.	
			1 sq.yard. = 0.8361 sq.metre.		
			1 sq.ft. = 0.0929 sq.metre.		
		1 sq.in.		= 6.4516 sq.cm.	
	1 hectare = 100 m × 100 m = 10000 sq. metre		= 2.4710 acres		
	4047 sq. metre		= 1 acre=4840 sq. yards		
Volume	1 cubic meter = 10^6 c.cm = 1000 litres		= 35.3357 cubic feet = 1.308 cubic yard		
			1 cubic inch = 16.387 cubic centimetres		
	1 cubic meter	= 219.969 UK gallons	= 264.172 US gallons		

	1 imperial gallon = 1.20095 US. Gallons = 4.545 litres				
	1 US gallon = 0.83267 Imperial gallon = 3.785 litres				
Volumetric flow	1 m^3/hr	=1000 litres/hour	= 4.4 gallons/ minute= 6336 gallons/day	16.67 litres/min	1 imperial gallon = 1.20095 US gallon
Mass	1 kg	=2.20462 Lb	=35.274 oz	$=10^{-3}$ tonne = 1.10231×10^{-3} short tonnes	= 1000 grams
	0.4536 kg	1 Lb = 16 Oz (avoir) = 453.6 grams			1OZ (avoir) = 28.35 grams
	Metric tonne = 1000 kg	= 1.12 short tonnes			Precious metals weigh scale; 1 Troy Oz = 31.1 grams
	Short tonne = 2000 pounds = 0.90718 metric tonnes	= 0.89287 long tonnes			= 32000 ounces

	1 long tonne = 2240 pounds				
					1gram = 1000 mg = 5 carats or 1 carat = 200 mg
Time	1 second	1 hour = 60 minutes	1 hour = 3600 seconds	1 day = 24 hr/day × 3600 seconds/hr = 86,400 seconds	
Velocity	1 metre/second	3.6 km/hr	2.237 miles/hr		
			1 mile/hr = 1.4667 feet/second = 0.447 metre/second		
Tempera-ture	0^0 K	-273.15^{0}c	-459.67^{0}F	C/5 = (F-32)/9	
	273.15^{0}K	0^0C	32^{0}F		
Force	9.81 Neutons	1kp = 2.205 lbf.	1lbf = 4.45N = 0.454 kp		
Water density	1 kg/m^3	1 gram/c.c.	62.428 lb/ft^3		

Pressure	1 bar = 10^5 N/m^2			= 75 cm Hg = 750.062 mm Hg	
	1 atmosphere = 14.73 pounds/sq. inch (psi)		33.9 feet of water =	760.062 mm Hg = 76 cm. Hg = 29.92 inches Hg	
	1 feet head water	0.432 psi			
Power	1 HP = 745.7 watts/second	1 HP (UK)	= 1.0139 hk (metric)	641.2 kcal/h	76.04 ppm/s
		1 HP (USA) = 33000 foot-lbs/min = 550 foot-lbs/second	=1.014 HP (metric)	= 42.44 BTU/min	
	1 kilowatt = KW	=1.3405 HP			
	1 KWh (kilo watt hour)	1.3405 Hp hour			
		1 HP = 550 foot-lbs/second			
Energy, work	Joule J = 2.778*10^{-7} Kwh	0.2388 × 10^{-3} Kcal	0.7375 foot pound	= 0.102 ppm	
Heat		1 BTU = 0.252 kg-calories	1 BTU=777.9 foot-lbs	1 Btu = 2.928 × 10^{-4} KW-hr	

Multiplier	Prefix				
10^{9}	Giga G				
10^{6}	Mega, M				
10^{3}	Kilo, k				
10^{2}	Hecto, h				
10	Deca, da				
10^{-9}	Naño, n				
10^{-6}	Micro, u				
10^{-3}	Milli, m				
10^{-2}	Centi , c				
10^{-1}	Deci, d				

Information sources: *Handbook of Chemistry & Physics*; *Metallurgy of* Non *-ferrous* Metals; *Sandwik Rock Processing Manual*; *Conversion Factors for the* Engineer by Dorr Oliver (gratefully acknowledged).

Appendix 6: Openings of Standard Test Sieves (US Standard & Tyler Standard Sieves)

	Openings of US Standard sieves			Openings of Tyler standard sieves				
Mesh size (2nd root of 2 in mm)	Inches (4th root of 2 in inches)	mm	Microns (1 mm = 1000 microns)	Mesh size	2nd root of 2 in Inches	4th root of 2 in inches	US series equivalent's Number	Microns (1 mm = 1000 microns)
3	¼" = 0.25	6.35	6350	3	0.263			
4	3/16" = 0.1875	4.76	4760	4	0.185		4	4760
5		4.0	4000	5		0.156	5	4000
6		3.36	3360	6	0.131		6	3360
7		2.83	2830	7		0.110	7	2830
8	3/32" = 0.0938	2.38	2380	8	0.093		8	2380
				9		0.078	10	2000
10		2.0	2000					
12			1680	10	0.065		12	1680
14			1410	12		0.055	14	1410
16			1190	14	0.046		16	1190
18			1000	16		0.039	18	1000
20			840	20	0.0328		20	840

25			710	24		0.0276	25	710
30			590	28	0.0232		30	590
35			500	32		0.0195	35	500
40			420	35	0.0164		40	420
45			350	42		0.0138	45	350
50			297	48	0.0116		50	297
60			250	60		0.0097	60	250
70			210	65	0.0082		70	210
80			177	80		0.0069	80	177
100			149	100	0.0058		100	149
120			125	115		0.0049	120	125
140			105	150	0.0041		140	105
170			88	170		0.0035	170	88
200			74	200	0.0029		200	74
230			62	250		0.0024	230	62
270			53	270	0.0021		270	53
325			44	325		0.0017	325	44
400			37	400	0.0015		400	37

Information source: *Handbook of Chemistry & Physics* (39th edition, 1957-1958).

References

1. Dennis,W.H. (1953). *Metallurgy of the nonferrous metals*. London: Pitman.
2. Gaudin, A.M. (1957). *Flotation* (2nd ed.). New York: McGraw Hill.
3. Taggart, A.F. (1951). *Elements of ore dressing* (1st ed.). New York: John Wiley.
4. *Handbook of chemistry and physics* (39th ed.). (1957-1958). Cleveland, OH: Chemical Rubber.
5. United States Steel Corporation. (1957). *The making, shaping, and treating of steel* (7th ed.). Pittsburgh, PA: United States Steel Corporation.
6. National Mineral Development Corporation Limited. (1989, February). *Performance guarantee test report for 11/c plant* (Bailadila Iron Ore Deposit No.14). Delhi: NDMC.
7. National Mineral Development Corporation Limited. (1992, August). *Beneficiation studies on iron ore fines from Siri Goa Mines of M/s Chowgule & Co., Ltd., Goa.* Hyderabad: R&D Centre, NMDC.
8. National Mineral Development Corporation Limited. (2012). *Making India a resource rich nation* (54th Annual Report, 2011-2012). Delhi: NMDC Ltd.
9. American Institute of Mining & Metallurgical Engineers. *Modern uses of nonferrous metals* (AIME Series, 2nd ed.). New York: AIME.
10. Indian Bureau of Mines. (2012, April). *Statistical profiles of minerals 2010-2011.* Chennai: Indian Bureau of Mines.
11. National Mineral Development Corporation Limited. (1992). *Separation of economic minerals out of beach sand from Bhimunipatnam*, Hyderabad.
12. Wills, B.A. & Napier-Munn, T. (2006). *Mineral processing technology (7th ed.). New York: Elsevier.*
13. *Sandvik rock processing.*
14. Denver Equipment Company.
15. Dana, E.S. (1922). *Dana's text book of mineralogy*. New York: John Wiley.
16. *Mining Engineer's Journal*, a publication of MEAI (Mining Engineers of India).
17. Experience gained/exposure to discussions with various firms, access to reports prepared as a part of consultancy assignment with MSPL Limited, Hospet, from February 1993 to March 2013.
18. Experience gained/exposure to discussions with various firms, access to reports prepared as a part of 12 short-term consultancy assignments with UNIDO, Jos, Nigeria, between June 1990 and August 1998.
19. Experience gained/exposure to discussions with various firms/organizations and access to several reports prepared while working with NMDC Ltd., Hyderabad, from November 1970 to November 1992.
20. Experience gained/exposure to discussions with various firms/organizations and access to several reports prepared while working with National Metallurgical Laboratory from October 1961 to November 1970.

21. Dr Gautam Dhar's presentation on demand and supply gap and coal security, made at the Associated Chamber of Commerce & Industry on 3 February 2011.
22. Ghosh, A. (2011, July 19). 'Nuclear boost: Uranium mine in Andhra could be amongst largest in the world', *TNN*.
23. Source: 1. 'Indian rare earths: Genesis & growth', T. K. Mukherjee, Indian Rare Earths Limited. 2. IREL Presentation by Dr. R. N. Patra.
24. 'Thorium reserves', by Sri Om Prakash Mathur.
25. 'Detailed information of tungsten ore in India', GSI, 1994.
26. Consultancy reports of METCHEM, Canada, on assignments entrusted by NMDC Ltd. and MSPL Ltd., separately.
27. Consultancy reports submitted by the author to UNIDO (Jos, Nigeria), and to MSPL Limited.
28. Read, H.H. (1970). *Rutley's elements of mineralogy*. London: T. Murby & Co.
29. Pictures of hematite, magnetite, copper, galena, sphalerite, gold & diamond from internet/Google.
30. Aerial views of Plant layout of Bailadila and Panna plants; Courtesy: National Mineral Development Corporation Limited, Hyderabad, 2014.

INDEX

Abbreviations used in this book

Symbol/Abbreviations	Used to denote/Description
#	Mesh size
ROM	Run of Mine ore / head
mm	Millimeters
+ "A" mm	Coarser than A mm size
-"A" + "B" mm	Smaller than "A" mm but coarser than "B" mm size
-"B" mm	Smaller than "B" mm size
-" A'' #	Finer than / passing through "A" mesh sieve/ screen
-200 mesh	Passing through 200 mesh sieve
BMQ	Banded Magnetite Quartzite
BHQ	Banded Hematite Quartzite
BHJ	Banded Hematite Jaspar
Taconite	Very low grade hard hematite/magnetite ore mostly available/mined in USA
Itaberite	Very low grade hard hematite ore mostly available/mined in Brazil
Wt %	Weight/yield percent
% Fe	Iron assay percent
% SiO_2	Silica assay percent
% Al_2O_3	Alumina assay percent
% recovery	Percent/extent of metal/element recovered in the product out of what is available in the ROM/Head
Lb/ton	Pounds per ton sample
Saleable product	Final product suitable for sale
Tails/ tailings	Rejects free from desired element/ metal/ mineral to be discarded into the tailing dump / dam for water recovery
NaCN	Sodium cyanide used for gold dissolution
Sp.gr	Specific gravity (number)
B.D	Bulk density in tonnes/cubic meter
Conc.	Concentrate
LIMS	Low Intensity Magnetic Separator
PRMS	Permanent Roll Magnetic Separator
DHIMS	Dry High Intensity Magnetic Separator
WHIMS	Wet High Intensity Magnetic Separator
M / mag.	Magnetic product
NM/non mag.	Non magnetic product
HMS	Heavy Media Separator
HTS	High Tension Separator

PPt	Precipitate
OF	Overflow
UF	Underflow
Head (actual / calc)	Metal/element/mineral content of ROM/test sample
Ro.	Rougher
Cl.	Cleaner
Recl.	Recleaner
Scav.	Scavenger
Midd.	Middlings containing values for further treatment
Ex works	Sale price at mine/ factory (to be lifted by the buyer at his cost)
PF	Power factor
% S	Percent solids in the pulp containing solids with water
M^3	Cubic meter
pH	Hydrogen ion concentration
BF	Blast furnace
DR	Direct reduction steel making furnace
DMS	Dense Media Separation
CSIR	Council of Scientific and Industrial Research
GMDC	Gujarat Mineral Development Corporation
GSI	Geological Survey of India
IBM	Indian Bureau of Mines
IREL	Indian Rare Earths Limited
KIOCL	Kudremukh Iron Ore Company
METCHEM	A Canadian Consulting firm engaged by KIOCL, NMDC etc. in India
MIT	Massachusetts Institute of Technology, USA
MOIL	Manganese Ores India Limited
MSPL	Mineral Sales Private Limited
NMDC	National Mineral Development Corporation Limited
NML	National Metallurgical Laboratory
OMM	Orissa Minerals & Mining
ONGC	Oil and Natural Gas Corporation Limited
R & D	Research and Development Centre
RRL	Regional Research Laboratory
SAIL	Steel Authority of India Limited
TISCO	Tata Iron and Steel Company
UNIDO	United Nations Industrial Development Organization

www.ingramcontent.com/pod-product-compliance
Ingram Content Group UK Ltd.
Pitfield, Milton Keynes, MK11 3LW, UK
UKHW041637190726
13854UKWH00006B/2538

9 789381 239544